HOW TO PASS

In full
COLOUR

STANDARD GRADE
MATHS

Ian Cassells
Mary Kay

HODDER
GIBSON
AN HACHETTE UK COMPANY

The Publishers would like to thank the following for permission to reproduce copyright material:

Photo credits

Page 80 © Richard Naude/Alamy

Acknowledgements

Formulae lists are reprinted by permission of the Scottish Qualifications Authority.

Every effort has been made to trace all copyright holders, but if any have been inadvertently overlooked the Publishers will be pleased to make the necessary arrangements at the first opportunity.

Although every effort has been made to ensure that website addresses are correct at time of going to press, Hodder Gibson cannot be held responsible for the content of any website mentioned in this book. It is sometimes possible to find a relocated web page by typing in the address of the home page for a website in the URL window of your browser.

Hachette's policy is to use papers that are natural, renewable and recyclable products and made from wood grown in sustainable forests. The logging and manufacturing processes are expected to conform to the environmental regulations of the country of origin.

Orders: please contact Bookpoint Ltd, 130 Milton Park, Abingdon, Oxon OX14 4SB. Telephone: (44) 01235 827720. Fax: (44) 01235 400454. Lines are open 9.00–5.00, Monday to Saturday, with a 24-hour message answering service. Visit our website at www.hoddereducation.co.uk. Hodder Gibson can be contacted direct on: Tel: 0141 848 1609; Fax: 0141 889 6315; email: hoddergibson@hodder.co.uk

© Ian Cassells and Mary Kay 2006, 2008
First published in 2006 by
Hodder Gibson, an imprint of Hodder Education,
an Hachette UK Company
2a Christie Street
Paisley PA1 1NB

This colour edition first published 2008

Impression number   5   4   3
Year   2012   2011   2010

Cover photo by Eric Heller/Science Photo Library

Illustrations by Tech-Set Ltd, Gateshead

Typeset in 10.5/14 Frutiger by Tech-Set Ltd, Gateshead

Printed and bound in Italy

A catalogue record for this title is available from the British Library.

ISBN-13: 978-0-340-97398 1

CONTENTS

CONTENTS

INTRODUCTION

Welcome to 'How to Pass Standard Grade Mathematics.'

By the time you are reading this, you may well be on the way to completing your Standard Grade course.

Over the past two years you will have been building up your mathematical knowledge and skills. Now the time has come to practise exam-type questions, using everything you have learned.

The good news is that the more you practise for the Standard Grade exam, the easier it becomes.

So let's get started.

In this book you will find:

◆ topics in both General and Credit Mathematics
◆ worked examples with explanations and tips
◆ more examples for you to try
◆ tips on exam techniques which could improve your exam mark.

If you are sitting the Credit exam, all of the questions and examples in this book will be applicable to you.

If you are not sitting the Credit exam, you should omit the parts marked with the Credit icon ©.

Prepare for the Big Day!

For effective revision, you should begin by getting yourself properly organised.

Make sure you have:

◆ 'How to Pass Standard Grade Mathematics' book
◆ class notes, summaries and jotters
◆ paper, pencils, ruler, rubber, protractor and a set of compasses
◆ scientific calculator (which you know how to use!).

Got everything? Well done.

(Keep all your books and equipment together during the revision period before your exam. A few minutes spent on organisation can save you time which you can then spend on your revision.)

The Examination Papers

Be familiar with the arrangements for the exams **you** are sitting. The lengths of the papers are as follows.

	Paper 1	Paper 2
General	35 minutes	55 minutes
Credit	55 minutes	1 hour 20 minutes

Formulae Lists

The **General** formulae list, printed on page 2 of the General paper, is as follows:

Circumference of a circle:	$C = \pi d$
Area of a circle:	$A = \pi r^2$
Curved surface area of a cylinder:	$A = 2\pi rh$
Volume of a cylinder:	$V = \pi r^2 h$
Volume of a triangular prism:	$V = Ah$

Theorem of Pythagoras:

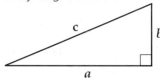

$$a^2 + b^2 = c^2$$

Trigonometric ratios in a right angled triangle:

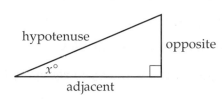

$$\tan x^\circ = \frac{\text{opposite}}{\text{adjacent}}$$

$$\sin x^\circ = \frac{\text{opposite}}{\text{hypotenuse}}$$

$$\cos x^\circ = \frac{\text{adjacent}}{\text{hypotenuse}}$$

Gradient:

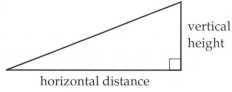

$$\text{Gradient} = \frac{\text{vertical height}}{\text{horizontal distance}}$$

The **Credit** formulae list, printed on page 2 of the Credit paper, is as follows:

The roots of $ax^2 + bx + c = 0$ are $x = \dfrac{-b \pm \sqrt{(b^2 - 4ac)}}{2a}$

Sine rule: $\dfrac{a}{\sin A} = \dfrac{b}{\sin B} = \dfrac{c}{\sin C}$

Cosine rule: $a^2 = b^2 + c^2 - 2bc \cos A$ or $\cos A = \dfrac{b^2 + c^2 - a^2}{2bc}$

Area of a triangle: $\text{Area} = \frac{1}{2}ab\sin C$

Standard deviation: $s = \sqrt{\dfrac{\sum(x-\bar{x})^2}{n-1}} = \sqrt{\dfrac{\sum x^2 - (\sum x)^2/n}{n-1}}$, where n is the sample size.

You need to know the General formulae when you sit the Credit exam. They are <u>not</u> printed in the Credit exam paper.

Know your Calculator!

During your revision work, use the calculator you will have in the exam.

Learn how **your** calculator works. (Believe it or not, calculators do vary.)

This will save you time during the exam and increase your confidence in Paper 2.

Remember, however, to show all your steps of working even when using your calculator.

Working

In every question, show clearly how you arrive at your answer. Marks are available for appropriate working, even if you do not complete the question.

Never score out working unless you have something better with which to replace it.

Let the Examiner decide if you have got anything correct.

What's It All About?

If you read a question and feel you really don't know how to tackle the problem, try this.

Read the question again and **underline** any words which are familiar to you, or words you feel might be important.

Think back to being in class when you were working on questions involving the words underlined. Try then to apply the knowledge and skills you used in class to the exam question.

You may not think you are going to complete the problem, but you may gain partial marks, or manage to jog your memory to let you complete the question!

Remember, the Examiner can only award you marks if you show appropriate working.

What Not to Do!

When studying for your Standard Grade exams:

◆ Do not merely read through your notes jotter, textbook, or this book **without doing something active**, like trying some examples. Otherwise, your mind may wander and not take in very much.

◆ Do not revise with the TV on or beside friends and family. This can be very distracting. Try to find a quiet place to study, away from distractions.

◆ Do not study for hours on end without a break. This may look good to your parents, but you won't take in much after the first hour. You should always take a 10-minute break from studying every hour.

◆ Beware of copying out notes from your jotter or textbook as a revision activity. This may not be a good idea as you can do it with your brain almost switched off. You will end up with a lot of paper on your desk but perhaps nothing much in your memory!

Chapter 2

HOW TO DEAL WITH NUMBERS

Numbers are part of our everyday lives. How many shopping days are there till Christmas? How much memory space is left on this disk? What area of grass do I need for this new lawn? What percentage of club members are children? How long is it till the next train comes?

When your employer and co-workers know that you have done Standard Grade Mathematics, they will expect you to be able to deal with all sorts of number work without any difficulty!

What You Should Know

a) for General Level: how to
  ◆ add and subtract (without a calculator) whole numbers and decimals to three decimal places
  ◆ multiply and divide (without a calculator) whole numbers and decimals to three decimal places, by a single digit or by multiples of 10, 100, 1000
  ◆ deal with all whole numbers and decimals (with a calculator) in applications of number, money and measure.

b) and in addition, for Credit Level:
  ◆ how to work with all whole numbers and decimals without a calculator.

Key Words

Calculate, **Sum**, **Product**, **Quotient**, and **Factor** are some of the words which can be used in questions involving number. (Do you know what these words mean?)

How to Add, Subtract, Multiply and Divide Whole Numbers and Decimals

The questions at the start of both the General and Credit non-calculator papers will give you the opportunity to use some of your basic numeracy skills. Question 1 in Paper 1 (non-calculator) at General Level has four short number questions. The first three parts are on basic number skills. Here are some of the different types of questions that can feature.

Example

Carry out the following calculations:

1 $16 \cdot 84 - 5 \cdot 3 + 3 \cdot 21$ **2** $43 \cdot 5 \times 8$ **3** $0 \cdot 592 \div 4$
4 $5 \cdot 38 - 4 \cdot 632$ **5** $7 \cdot 58 \times 30$ **6** $1800 \div 60$
7 $2850 \div 5000$ **8** $4 \cdot 3 \times 700$.

(Solution)

1
$$\begin{array}{r} 16 \cdot 84 \\ - 5 \cdot 30 \\ \hline 11 \cdot 54 \end{array} \qquad \begin{array}{r} 11 \cdot 54 \\ + 3 \cdot 21 \\ \hline 14 \cdot 75 \end{array}$$
(insert a zero)

2
$$\begin{array}{r} 43 \cdot 5 \\ \times 8 \\ \hline 348 \cdot 0 \end{array}$$

3
$$4 \overline{)0 \cdot 592} \quad 0 \cdot 148$$

4 $5 \cdot 380$ ◄──(insert a zero)
$$\begin{array}{r} 5 \cdot 380 \\ - 4 \cdot 632 \\ \hline 0 \cdot 748 \end{array}$$

5 $7 \cdot 58 \times 10 = 75 \cdot 8$
$75 \cdot 8 \times 3 = 227 \cdot 4$

6
$$60 \overline{)1800} \quad 30$$

7 $2850 \div 1000 = 2 \cdot 85$
$2 \cdot 85 \div 5 = 0 \cdot 57$

8 $4 \cdot 3 \times 100 = 430$
$430 \times 7 = 3010$.

The first two or three questions in the Credit Paper 1 are likely to test you on adding, subtracting, multiplying and dividing whole numbers and decimals, as well as fractions and percentages.

Example

Evaluate:
$4 \cdot 7 - (5 \cdot 52 \div 6)$.

(Solution) $4 \cdot 7 - (5 \cdot 52 \div 6)$
$= 4 \cdot 7 - (0 \cdot 92)$
$= 3 \cdot 78$.

There are usually between ten and thirteen questions in Paper 1 for both General and Credit, so your non-calculator skills have to be good.

In Paper 2 for both General and Credit, you are expected to be able to use your calculator to do calculations with whole numbers and decimals. It is therefore important that you know how to key the mathematical operations required into your calculator.

The number work in the Credit exam includes the use of brackets too.

For Practice

Using your calculator:

◆ Can you find squares and square roots (e.g. $\sqrt{7}$)?

◆ What about cubes like 5^3?

◆ Can you use the π button?

◆ Can you use sin, cos and tan successfully?

◆ Do you know how to 'store' a number in the memory and 'recall' it later?

◆ Do you know how to set your calculator so that it calculates in degrees? (It may have been changed by accident.)

◆ Can you make use of the 'ANS' button?

(If you can do all of the above, you know your calculator well enough for Standard Grade Mathematics.)

Common Mistakes

1 When subtracting decimals e.g. $8 \cdot 6 - 4 \cdot 73$ a common wrong answer is $3 \cdot 93$. It may have been arrived at as follows.

$$\begin{array}{r} 8 \cdot 6 \\ - 4 \cdot 73 \\ \hline 3 \cdot 93 \end{array}$$

Don't forget here that $8 \cdot 6$ needs to written as $8 \cdot 60$.
(correct answer)

$$\begin{array}{r} 8 \cdot 60 \\ - 4 \cdot 73 \\ \hline 3 \cdot 87 \end{array}$$

2 In a question which tests adding and subtracting decimals, **do not** swap the plus and minus signs around.

For example, $5 \cdot 63 - 7 \cdot 8 + 4 \cdot 16$ **does not** become $5 \cdot 63 + 7 \cdot 8 - 4 \cdot 16$ just because it makes the question easier.

(correct answer) $5 \cdot 63 + 4 \cdot 16 = 9 \cdot 79$, leading to $9 \cdot 79 - 7 \cdot 8 = 1 \cdot 99$.

Key Points

◆ Get plenty of practice in the type of questions that appear at the start of the General and Credit non-calculator papers.

◆ When adding or subtracting decimals, make sure the decimal points line up as follows:

$$
\begin{array}{r}
23{\cdot}45 \\
+\,54{\cdot}28 \\
\hline
77{\cdot}73
\end{array}
$$

◆ When multiplying decimals remember that there are as many numbers after the decimal point in the answer as there were **in total** after the points in the question. Thus:

$$7{\cdot}18 \times 3 \qquad \text{(two numbers after the point in the question)}$$
$$= 21{\cdot}54 \qquad \text{(two numbers after the point in the answer)}$$

◆ Multiplying and dividing decimals by 100, 1000 etc. means that you move the numbers past the point by the number of zeros e.g. $2{\cdot}34 \times 1000 = 2340$.

$$
\begin{array}{r}
2{\cdot}34 \\
\Rightarrow \quad 23{\cdot}4 \\
\Rightarrow \quad 234{\cdot}0 \\
\Rightarrow \quad 2340{\cdot}0
\end{array}
$$

◆ Get the best out of your calculator by learning to use all the functions you will require for your exam.

Chapter 3

PERCENTAGES, FRACTIONS AND INTEGERS

Percentages, Fractions and Integers occur in our everyday lives. We all know that banks pay interest when we save with them; interest is always advertised as a percentage rate. We often mention common fractions in everyday conversation, 'I could only eat half a pizza', or 'I'm meeting my mum at a quarter past four'. In the winter weather forecasters use integers to warn of freezing temperatures like $-8°C$. People who are comfortable with percentages, fractions and integers can deal with more complex problems such as being able to work out the percentage increase in a business profit, or calculating the final value of something which is decreasing in value at the rate of 12% each year for 4 years, or being able to calculate the fraction of voters who voted Labour at the last election.

Key Words and Definitions

A **Percentage** is usually a whole number representing a fraction of 100. (Percentages are often used in calculating wages, savings accounts, interest, and VAT.) A **Loan** is money that is borrowed and usually paid back with interest. **Interest** is the amount charged to the borrower for using the lender's money. **Per Annum** is Latin for 'each year'. In **Fractions**, the **Numerator** is the upper number and the **Denominator** is the lower number. Fractions are called **top-heavy fractions** when the number on top is bigger than the number on the bottom. **Integers** are positive or negative whole numbers.

What You Should Know

a) for General Level: how to
   ◆ find percentages of numbers and quantities
   ◆ write one quantity as a percentage of the other
   ◆ add, subtract, and multiply simple fractions
   ◆ find simple fractions of quantities
   ◆ relate decimals, fractions and percentages, one to another
   ◆ add or subtract integers
   ◆ multiply a single digit integer by a single digit whole number (e.g. $3 \times (-4)$).

b) and in addition, for Credit Level: how to
   ◆ add, subtract, multiply and divide all fractions including mixed numbers.

HOW TO PASS STANDARD GRADE MATHEMATICS

Finding Percentages: Percentage Increases; Percentage Reductions.

Example

Find 35% of 84 kilograms.
(This might appear in a non-calculator paper.)

(Solution)

10% of 84 = 8·4
so 30% = 8·4 × 3 = 25·2
also 5% of 84 = 4·2 (since 5% = $\frac{1}{2}$ of 10%)
hence 35% of 84 kg = 25·2 + 4.2 = 29·4 kg.

Example

A camera, which originally cost £150, has been reduced in price by £20.
Calculate the percentage reduction.
(This type of question might appear in a calculator paper.)

(Solution)

$$\text{Percentage Reduction} = \frac{\text{reduction}}{\text{original cost}} \times 100$$

$$= \frac{20}{150} \times 100$$

$$= 13 \cdot 3\% \text{ (to 1 decimal place).}$$

Fractions: The Basic Operations

The topic of fractions is one that most students of mathematics dislike. However, you are likely to meet fractions at least once in both the General and Credit exams. To help you feel more confident about fractions read through the problems below, then try the student examples at the end. Hopefully you should then be able to tackle questions involving fractions successfully in the SQA past papers.

Example

A school has 750 pupils; $\frac{3}{5}$ of them walk to school.
How many pupils walk to school?

(Solution)

$$\frac{1}{5} \times 750 = 150$$

$$\Rightarrow \frac{3}{5} \times 750 = 450 \text{ pupils}$$

Hence 450 pupils walk to school.

Example

a) Calculate $\dfrac{1}{4} + \dfrac{5}{8}$

(Solution) $\dfrac{1}{4} + \dfrac{5}{8} = \dfrac{2}{8} + \dfrac{5}{8}$ *(make the denominators the same)*

$\qquad\qquad = \dfrac{7}{8}$ *(add the numerators)*
(do not add the denominators!)

b) Calculate $\dfrac{7}{12} - \dfrac{1}{3}$

(Solution) $\dfrac{7}{12} - \dfrac{1}{3} = \dfrac{7}{12} - \dfrac{4}{12}$ *(make the denominators the same)*

$\qquad\qquad = \dfrac{3}{12}$ *(subtract the numerators)*

$\qquad\qquad = \dfrac{1}{4}$ *(divide both numerator and denominator by 3)*

c) Calculate $2\dfrac{2}{3} + 3\dfrac{5}{6}$

(Solution) $2\dfrac{2}{3} + 3\dfrac{5}{6} = 5\dfrac{2}{3} + \dfrac{5}{6}$ *(first add the whole numbers)*

$\qquad\qquad = 5\dfrac{4}{6} + \dfrac{5}{6}$ *(use a common denominator)*

$\qquad\qquad = 5\dfrac{9}{6}$ *(add the numerators)*

$\qquad\qquad = 6\dfrac{3}{6}$ *(simplify $\dfrac{9}{6}$ to $1\dfrac{3}{6}$)*

$\qquad\qquad = 6\dfrac{1}{2}$ *(simplify by dividing numerator and denominator by 3)*

d) Calculate $\dfrac{3}{4} \times \dfrac{5}{7}$

(Solution) $\dfrac{3}{4} \times \dfrac{5}{7} = \dfrac{3 \times 5}{4 \times 7}$ *(multiply numerators together and multiply denominators together)*

$\qquad\qquad = \dfrac{15}{28}$

Example

a) Calculate $\dfrac{1}{3} \times 1\dfrac{5}{7}$

(Solution) $\dfrac{1}{3} \times 1\dfrac{5}{7} = \dfrac{1}{3} \times \dfrac{12}{7} = \dfrac{12}{21}$ *(first change mixed number to a vulgar fraction)*

$= \dfrac{4}{7}$ *(divide top and bottom by 3)*

b) Calculate $1\dfrac{2}{3} \div \dfrac{5}{8}$

(Solution) $1\dfrac{2}{3} \div \dfrac{5}{8} = \dfrac{5}{3} \div \dfrac{5}{8}$ *(first change the mixed number to a vulgar fraction)*

$= \dfrac{5}{3} \times \dfrac{8}{5} = \dfrac{40}{15}$ *(change the ÷ to × and turn 2nd fraction upside down)*

$= \dfrac{8}{3}$ *(divide top and bottom by 5)*

$= 2\dfrac{2}{3}$ *(change vulgar fraction to mixed number)*

Here are a few questions with summary solutions which are similar to those at the start of the Credit non-calculator paper:

Example

Calculate:

a) $7\cdot82 - 1\cdot6 \times 3$

b) $2\dfrac{5}{6} \div \dfrac{2}{3}$

c) $7\cdot8 - (5\cdot07 - 2\cdot1)$

d) $\dfrac{2}{5}\left(1\dfrac{2}{3} + \dfrac{5}{6}\right).$

(Solutions)

a) $7\cdot82 - 4\cdot8 = 3\cdot02$

b) $\dfrac{17}{6} \div \dfrac{2}{3} = \dfrac{17}{6} \times \dfrac{3}{2} = \dfrac{17}{2} \times \dfrac{1}{2} = \dfrac{17}{4} = 4\dfrac{1}{4}$

c) $7\cdot8 - 2\cdot97 = 4\cdot83$

d) $\dfrac{2}{5}\left(1\dfrac{4}{6} + \dfrac{5}{6}\right) = \dfrac{2}{5}\left(2\dfrac{1}{2}\right) = \dfrac{2}{5} \times \dfrac{5}{2} = 1.$

Converting Fractions, Decimals, and Percentages

Having to change a fraction to a decimal or a decimal to a fraction isn't something you have to do very often in Standard Grade Maths, but it is very important that you know how to perform both of these tasks when required.

Example

Write $\frac{5}{6}$ as a decimal. (Give your answer correct to two decimal places.)

(Solution) $\frac{5}{6}$ can be thought of as 5 divided by 6.

$$5 \div 6 = 6\overline{)5 \cdot 0}$$

$$= 6\overline{)5 \cdot 000}^{\,0 \cdot 8333}$$

$$= 0 \cdot 83 \text{ (to 2 decimal places).}$$

Being able to compare fractions, decimals and percentages is useful for questions like the one which follows.

Example

Starting with the smallest write the following in order:

$$0 \cdot 303 \qquad \frac{1}{3} \qquad 31\% \qquad 0 \cdot 03 \qquad \frac{3}{10}$$

(Solution)   If you change all the above numbers to decimals they are easier to compare, and hence to put in order.

$$1 \div 3 = 0 \cdot 333$$

$$31\% = \frac{31}{100} = 0 \cdot 31$$

$$\frac{3}{10} = 0 \cdot 3$$

So the correct order is $0 \cdot 03$, $0 \cdot 3$, $0 \cdot 303$, $0 \cdot 31$, $0 \cdot 333$ and

final solution $= 0 \cdot 03$, $\frac{3}{10}$, $0 \cdot 303$, 31%, $\frac{1}{3}$.

HOW TO PASS STANDARD GRADE MATHEMATICS

Working with Negative Integers

Negative Integers (whole numbers) feature regularly in the General paper, usually in a real life setting.

Example

The temperature recorded at 4 pm in Glasgow one day last winter was 5 °C.
By 6 am the following morning, the temperature had fallen to −3 °C.
By how many degrees had the temperature fallen?

(Solution)   The fall in temperature between 5 °C and 0 °C is 5°,
and the fall in temperature between 0 °C and −3 °C is 3°.
So the temperature had fallen a total of 8 °C.

Sometimes you will have to add, subtract, multiply or divide negative numbers or a combination of these.

Example

The operation ◊ means 'multiply the first number by 3 and add the second number'.
(For example $-2 \lozenge 5 = 3 \times (-2) + 5 = -6 + 5 = -1$.)

a) Calculate $-5 \lozenge 7$.   b) If $p \lozenge 6 = -9$, find p.

(Solution)   a) $3 \times (-5) + 7 = -15 + 7 = -8$.

b)
$$p \lozenge 6 = -9$$
$$\Rightarrow \quad 3 \times p + 6 = -9$$
$$\Rightarrow \quad 3p + 6 = -9 \qquad \textit{(an equation in p)}$$
$$\Rightarrow \quad 3p = -15 \qquad \textit{(subtract 6 from each side)}$$
$$\Rightarrow \quad p = -5. \qquad \textit{(divide each side by 3)}$$

Being able to square or cube a negative number is necessary for Credit.
For example, $(-3)^2 = 9$ and $(-4)^3 = -64$.

Example

Subtract −8 from 23.

(Solution) $23 - (-8)$
$= 23 + 8 \qquad \textit{(subtracting a negative is the same as adding)}$
$= 31$.

Multiplying and Dividing Integers

I'm sure you were familiar with multiplying and dividing integers in first or second year. If you have forgotten what happens when you multiply or divide by a negative number then read the following very carefully.

Remember

Multiplying two positive numbers together gives a positive answer.
Multiplying a positive number by a negative number gives a negative answer.
Multiplying two negative numbers together gives a positive answer.
Hence:

$3 \times 5 = 15$	(+ve number multiplied by +ve number = +ve number)
$3 \times (-5) = -15$	(+ve number multiplied by −ve number = −ve number)
$-3 \times 5 = -15$	(−ve number multiplied by +ve number = −ve number)
$-3 \times (-5) = 15.$	(−ve number multiplied by −ve number = +ve number)

The rules above can be simplified by using the table below.

×	+	−
+	+	−
−	−	+

Key Points

◆ Learn the equivalence of well-known fractions such as $25\% = \dfrac{1}{4} = 0.25$.

◆ Use the fact that $10\% = \dfrac{1}{10}$, $5\% = \dfrac{1}{2}$ of 10%, and $1\% = \dfrac{1}{100}$ to help you calculate percentages of quantities quickly.

◆ Learn the formulae for percentage increase and percentage reduction.

◆ To find a fraction of a quantity, divide the quantity by the denominator (bottom) and multiply by the numerator (top).

◆ When adding or subtracting fractions make the denominators the same.

◆ When multiplying fractions change any mixed numbers such as $2\dfrac{1}{2}$ into vulgar (top heavy) fractions $\left(\dfrac{5}{2}\right)$.

◆ To change from a fraction to a decimal, divide the numerator (top) by the denominator (bottom).

◆ Adding a negative number is the same as subtracting $(5 + (-3) = 5 - 3 = 2)$.
Subtracting a negative number is the same as adding $(5 - (-3) = 5 + 3 = 8)$.

Finding the percentage or fraction of a quantity together with calculations involving negative numbers are often found in questions in the General non-calculator paper. As well as trying the questions below, you should look at the recent SQA past papers and identify the questions in which you might have to use Fractions, Percentages and Integers. In fact, it might be a good idea to try some of the SQA past paper questions now.

For Practice

(The answers to the following questions are given in Appendix 2.)

3.1 Find 45% of 76 litres.

3.2 A kitchen table which originally cost £269 has been increased in price to £299. Calculate the percentage increase.

3.3 A survey of 1400 adults shows that $\frac{3}{4}$ of them own mobile telephones. How many adults own a mobile telephone?

3.4 Calculate $\frac{1}{6} + \frac{2}{3}$.

3.5 Calculate $\frac{3}{4} - \frac{5}{16}$.

3.6 Calculate $3\frac{5}{7} - 2\frac{1}{3}$.

3.7 Calculate $\frac{5}{6} \times \frac{3}{4}$.

3.8 Calculate $2\frac{1}{4} \times 1\frac{3}{5}$.

3.9 Write $\frac{5}{7}$ as a decimal.

3.10 Starting with the biggest write the following in order:

    0·71 $\frac{5}{7}$ 73%   0·707 $\frac{7}{10}$.

3.11 The weather station at the top of Cairn Gorm mountain recorded a temperature of −12 °C one night last winter. The following day the temperature rose to 7 °C. By how many degrees had the temperature risen?

3.12 The operation ▲ means 'multiply the first number by 4 and subtract the second number'.
Calculate −2 ▲ 5.

3.13 Subtract −15 from 6.

3.14 Evaluate $y = x^2 + 2x$, when $x = -3$.

3.15 Evaluate $m = \frac{n^2 - 3n}{2n}$, when $n = -1$

Chapter 4

NUMBERS LARGE AND SMALL

The number of seconds in a century is 3 155 760 000.
The wavelength of violet light is 0·000 000 41 metre.

Fortunately very large and very small numbers like these can be written in a shorthand way, called **Scientific Notation**. A number written in scientific notation is a decimal number (greater than or equal to 1 and less than 10) multiplied by 10 raised to a power.

Key Points

For example, the large number 57 600 can be written in scientific notation as $5·76 \times 10^4$.

The rules for doing this are:
◆ move the numbers to the right until the decimal point is between 5 and 7 …5·76
◆ count how many places you moved the numbers to the right, …4
◆ write this number as your power of 10, …10^4.

The process for changing small numbers to scientific notation is slightly different. For example, the number 0·00318 can be written in scientific notation as $3·18 \times 10^{-3}$.
The rules for doing this are:
◆ move the numbers to the left until the decimal point is between the 3 and 1 (after the first non-zero digit) …3·18
◆ count how many places you moved the numbers to the left, …3
◆ write this number as your negative power of 10, …10^{-3}.

What You Should Know

a) for General Level:
◆ how to write in scientific notation for large and small numbers using whole number powers only (for example, 7×10^3, $3·28 \times 10^1$, $6·4 \times 10^{-2}$, $a \times 10^n$)
◆ how to write a number given in scientific notation as a full number (7×10^3 as 7000).

b) and in addition, for Credit Level:
◆ how to multiply and divide numbers in scientific notation.

HOW TO PASS STANDARD GRADE MATHEMATICS

Key Words

Power, **Index**, **Square Root** and **Cube Root** are some of the words which can be used in questions involving index notation. (Do you know what these words mean?)

Example

The area of Canada is approximately $9{\cdot}98 \times 10^6$ square kilometres. Write this number in full.

(Solution)   9 980 000 square kilometres.

(Here the number 9·98 has been moved six places (since 10^6) to the left.)

Very Small Numbers given in Scientific Notation sometimes have to be written in full too.

Example

A typical human cell is $1{\cdot}1 \times 10^{-2}$ millimetres in diameter. Write this number in full.

(Solution)   0·011 millimetres.

(Here the number 1·1 has been moved two places (since 10^{-2}) to the right.)

Remember

$10^1 = 10$	$10^{-1} = 0{\cdot}1$
$10^2 = 100$	$10^{-2} = 0{\cdot}01$
$10^3 = 1000$	$10^{-3} = 0{\cdot}001$
$10^4 = 10\ 000$	$10^{-4} = 0{\cdot}0001$
$\vdots \qquad \vdots$	$\vdots$
$10^6 = 1\ 000\ 000$	$10^{-6} = 0{\cdot}000\ 001$
$\vdots \qquad \vdots$	$\vdots$
$10^9 = 1\ 000\ 000\ 000$	$10^{-9} = 0{\cdot}000\ 000\ 001$

When **multiplying** numbers in scientific notation (without using a calculator) remember to **add** the powers of ten. When you are **dividing** you subtract one power of ten from the other. (See the later section on Indices for more work on powers.)

Example

Calculate $(7\cdot3 \times 10^3) \times (4 \times 10^4)$.

(Solution)   First multiply 7·3 by 4 = 29·2
Then add the powers of 10^3 and $10^4 = 10^7$
Put the two answers together to give $29\cdot2 \times 10^7$
In proper Scientific Notation, $29\cdot2 \times 10^7$ then becomes $2\cdot92 \times 10^8$.

Calculate $(5\cdot68 \times 10^2) \div (8 \times 10^5)$.

(Solution)   First divide 5·68 by 8 = 0·71
Then subtract the powers of 10^2 and $10^5 = 10^{-3}$
Put the two answers together to give $0\cdot71 \times 10^{-3}$
In proper Scientific Notation, $0\cdot71 \times 10^{-3}$ then becomes $7\cdot1 \times 10^{-4}$.

To perform calculations such as $7 \times (3\cdot94 \times 10^3)$ on your calculator you need to know how to use the (EXP) or (EE) button. Thus:

Enter: $7 \times 3\cdot94$ (EXP) 3 =

This will give you an answer of 27 580 or $2\cdot758 \times 10^4$.

Example

Last year a total of $8\cdot2 \times 10^6$ people flew into or out of Glasgow Airport.
How many people was this per day?
(Give your answer in scientific notation.)

(Solution) $8\cdot2 \times 10^6 \div 365$ ($\Rightarrow$ Enter 8·2 (EXP) 6 ÷ 365)
= 22 466
= $2\cdot25 \times 10^4$ (rounded to two decimal places – see the later section
on Rounding).

(In the Credit calculator paper you may be asked a question like the following one.)

Example

In the Physics formula $E = \frac{1}{2} mv^2$, E is the Kinetic Energy of a body of mass m, which is travelling with velocity v.

Find the value of E when $m = 8{\cdot}1 \times 10^{-2}$ and $v = 3 \times 10^7$.

(Write your answer in scientific notation.)

(Solution)
$$E = \frac{1}{2} \times (8{\cdot}1 \times 10^{-2}) \times (3 \times 10^7)^2 = \frac{1}{2} \times (8{\cdot}1 \times 10^{-2}) \times (3 \times 10^7) \times (3 \times 10^7)$$

The display of a scientific calculator might show 36450000000000 or $3{\cdot}645 \times 10^{13}$.

Both answers are the same and written as $3{\cdot}645 \times 10^{13}$ in scientific notation.

Common Mistakes

Common mistakes in Scientific Notation questions include, for example, writing 23 400 as $2{\cdot}34^4$ ($2{\cdot}34 \times 10^4$ is correct). Another error is to write 23 400 as 234×10^2, or sometimes as $23{\cdot}4 \times 10^3$. It is easy to count the powers of 10 wrongly, therefore you should perhaps double check your powers of ten.

Summary

◆ Scientific Notation is the name used for the shorthand way of writing very large or very small numbers.
◆ Large numbers (e.g. 580 000) or small numbers (e.g. 0·000 056) can both be changed into Scientific Notation.
◆ Numbers written in Scientific Notation can also be re-written in full as large or small numbers.

For Practice

(The answers to the following questions are given in Appendix 2.)
4.1   The area of Brazil is approximately $8{\cdot}51 \times 10^6$ square kilometres.
      Write this number in full.

For Practice continued ➢

For practice *continued*

4.2 Microsoft's turnover in the year ending June 2004 was $3{\cdot}68 \times 10^{10}$.
How much was this per month?
Give your answer in scientific notation.

4.3 Calculate $(3{\cdot}46 \times 10^4) \times (6{\cdot}71 \times 10^2)$.

4.4 Calculate $(8{\cdot}47 \times 10^7) \div (2{\cdot}32 \times 10^3)$.

4.5 $G = a^2 + ab$
Find the value of G when $a = 2{\cdot}17 \times 10^6$ and $b = 6{\cdot}25 \times 10^{-9}$.
Give your answer in scientific notation.

NUMBERS LARGE AND SMALL

RATIO, PROPORTION AND VARIATION

i

A knowledge of ratio, proportion and variation allows us to perform calculations involving simple arithmetic relationships.

What You Should Know

a) for General Level: you should be familiar with
◆ how to split a quantity in a given ratio e.g. $4:1$, $1:3$
◆ direct and inverse proportion
◆ direct variation and its graph
◆ use of scale drawings to solve problems
◆ how to work out a scale factor from a diagram
◆ ratio of sides of similar right-angled triangles
◆ ratio of areas of similar figures.

b) and in addition, for Credit Level: familiar with
◆ how to split a quantity in a given ratio e.g. $3:5$
◆ inverse variation and its graph; joint variation
◆ using ratio or scale factor to calculate the lengths of sides of similar triangles
◆ ratio of surface areas of similar solids
◆ ratio of volumes of similar solids.

Working with Ratio

A 'ratio' is a numerical comparison of two quantities. For example, you can look at a group of pupils and count the ratio of boys to girls in the group. If the group has twenty-five pupils, and ten of those are girls, then the ratio of girls to boys is 10 to 15. The ratio of boys to girls is $15:10$ however.

A '10 to 15' ratio can also be written as $10 : 15$ or as $\frac{10}{15}$. It can also be simplified to $2:3$ or $\frac{2}{3}$ (by dividing throughout by 5).

Hints *and* Tips

To simplify a ratio, divide both parts by the largest number that can be divided into both numbers without remainder. To simplify $16:20$ you would divide by 4 and the simplified ratio would be $4:5$.

Example

The ratio of children to adults in the audience at a school show was $1:4$.
The audience total was 195.
How many adults were in the audience at the show?

(Solution) $1 + 4 = 5$ 'shares'   *(add up the total number of 'shares')*
                   $\Rightarrow$ 5 shares $= 195$
                   $\Rightarrow$ 1 share $= 195 \div 5 = 39$ *(divide by 5 to find 1 'share')*
                   $\Rightarrow$ 4 shares $= 39 \times 4 = 156$ *(multiply by 4 to find 4 'shares')*

             There were 156 adults in the audience.

Hints *and* Tips

When dividing a total in a ratio it helps to add up the number of possible 'shares'. For example, if £210 is to be shared between two people in the ratio $4:3$, there would be 7 shares; if the ratio were $1:2$ there would be 3 shares; and for a ratio of $7:3$ there would be 10 shares. You can then work out how much money each person gets by following the example above.

Example

Janie mixes red and blue paint in the ratio $2:5$ to make purple paint.
She has 10 litres of red and 15 litres of blue paint.
What is the maximum amount of purple paint Janie can make?

(Solution)
For every 2 litres of red Janie uses she needs 5 litres of blue.
The amount of blue paint she has restricts the amount of purple paint she can make.
She can use 5 litres of blue paint a maximum 3 times.
Hence red paint $= 3 \times 2$ litres $= 6$ litres
and blue paint $= 3 \times 5$ litres $= 15$ litres.

Janie can make $15 + 6 = 21$ litres of purple paint.

i

Ratio and Scale Drawing

An **Enlargement** is larger than an original shape and a **Reduction** is smaller. **Scale Factor** is a measure of the enlargement or reduction, for example, a scale factor of 3 (or 3 : 1) means three times larger, and a scale factor of $\frac{1}{2}$ (or 1 : 2) is better known as half size.

In scale drawings, the scale is given as a ratio, either in words such as 1 cm to 1 km, or more commonly in figures such as 1 : 500. You may be familiar with maps where the scale is 1 : 50 000.

Example

a) On a 1 : 50 000 map, two churches are 4 cm apart. What is the actual distance between them?

b) Two Post Offices are 12 km apart. How far apart will they appear on the map:

(Solution)

a) On the map, each centimetre represents 50 000 cm

    Hence 4 cm represents $4 \times 50\,000$ cm $= 200\,000$ cm

                                     $= 2000$ m

                                     $= 2$ km.

b) Since 2 km is represented by 4 cm,

    then 12 km is represented by 6×4 cm

    The Post Offices are therefore 24 cm apart on the map.

If you are asked to do a *Scale Drawing*, you will find it fairly straightforward if you use a step-by-step approach, as is used in the following example.

Example

The diagram shows the design of a blade for a craft knife.

Redraw this diagram to a scale of 3 : 1.

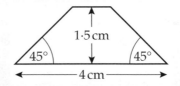

The scale in this question is 3 : 1. This means that the final drawing has to be three times bigger than the original, but with the same shape. The angles are not changed in any way!

We must first decide on a starting point, and then work systematically from that point.

i

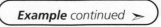

Example *continued* ➢

Example *continued*

(Solution)

From point O draw in the base line 12 cm long, (1).

With a protractor measure 45°, from the base line at O, (2).

Draw the line which is at 45° to the base, (3).

The top line can be drawn parallel to the base, 4·5 cm above it (4).

The final line of the diagram is drawn from the right hand end of the base at 45° to the base, (5). This completes the shape. Thus:

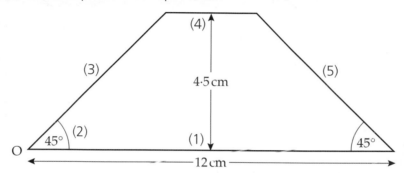

 ## *Similarity*

Similar Shapes

Two or more shapes are **similar** if they have the same shape, but are not necessarily the same size. Similar shapes have their corresponding sides in proportion. That is, the ratio of corresponding sides is the same. Their corresponding angles are equal.

Rectangles A and B are **similar**.

B is an enlargement of A – the scale factor of the enlargement is 3.

A is a reduction of B – the scale factor of the reduction is $\frac{1}{3}$.

Note, too, the equality of ratios: $1:2(A) = 3:6(B)$; also $3:1(B:A) = 6:2(B:A)$.

Key Points

◆ When a shape is enlarged or reduced the ratios of the lengths of the corresponding sides are the same. (The enlarged/reduced shape and the original are **similar**.)

◆ In an enlargement the scale factor is greater than 1.

◆ In a reduction the scale factor is less than 1.

Similar Triangles

When two or more triangles are equiangular (corresponding angles equal), their corresponding sides are in the same ratio and the triangles are similar. The following diagram shows two similar triangles.

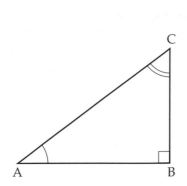
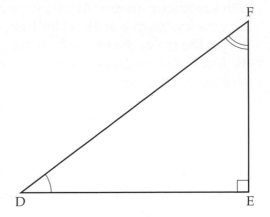

The corresponding angles have been marked. You may wish to check by measurement that $\angle CAB = \angle FDE$, and $\angle ACB = \angle DFE$.

Since the triangles are **equiangular**, they are **similar**.

Hence $AB:DE = BC:EF = AC:DF$, since the ratio of corresponding sides is the same.

(Also $AB:BC = DE:EF$, $AB:AC = DE:DF$, and $AC:BC = DF:EF$.)

Example

A radio mast 4 metres high is positioned as shown on top of a tower.

A steel wire joins the top of the mast to point P on the ground.

Calculate the height, h, of the tower.

(Solution)
Triangles PQR and RST are similar (why?).

Hence $\dfrac{h}{12} = \dfrac{4}{3}$

$\Rightarrow 3h = 48$ *(by cross multiplying)*

$\Rightarrow h = 16\,\text{m}.$

The height of the tower is thus 16 m.

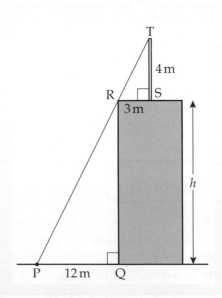

C Check that the following triangles are similar and write down and check equal ratios for corresponding sides. Similar triangles are not always right-angled!

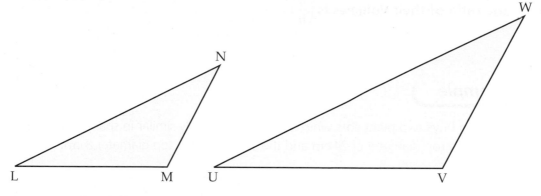

Similarity and Area

If two shapes are similar and the ratio of their corresponding sides is $\frac{a}{b}$, then the ratio of their Areas is $\left(\frac{a}{b}\right)^2$.

Example A tile manufacturer makes two sizes of tile in the shape of a regular hexagon. The smaller tile has an edge length of $10\,cm$, and the larger one has edge length $15\,cm$.

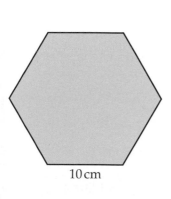

10 cm

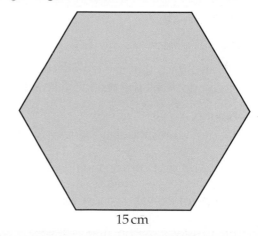

15 cm

The smaller tile has an area of $260\,cm^2$.
What is the area of the larger tile?

(Solution)

Ratio of corresponding sides $= \dfrac{15}{10} = \dfrac{3}{2}$.

Hence ratio of areas $= \left(\dfrac{3}{2}\right)^2 = \dfrac{9}{4}$.

Hence area of larger tile $= \dfrac{9}{4} \times 260 = 585\,cm^2$.

Similarity and Volume

If two objects are similar and the ratio of their corresponding sides is $\frac{a}{b}$, then the ratio of their Volumes is $\left(\frac{a}{b}\right)^3$.

Example

Maralyn buys two plant pots which are mathematically similar in shape. The large one has a top diameter of 24 cm and the small one has top diameter 8 cm.

Maralyn finds that she requires 0·3 litre of compost to fill the small pot. What volume of compost will she need to fill the large pot?

(Solution)
Ratio of corresponding 'sides' $= \frac{24}{8} = \frac{3}{1}$.

Hence ratio of corresponding volumes $= \left(\frac{3}{1}\right)^3 = 3^3 = 27$.

Hence volume of compost for larger pot $= 27 \times 0.3 = 8{\cdot}1$ litres.

Proportion

Direct Proportion

Two quantities are said to be in Direct Proportion when one is a multiple of the other. For example, the total weight of sugar is directly proportional to the number of bags. If one bag weighs 1 kg, then five bags weigh 5 kg, ten bags weigh 10 kg and so on.

Example

Gordon buys 12 fence posts for his garden at a cost of £54.
How much would he pay for another seven fence posts?

(Solution)

12 fence posts cost £54

$\Rightarrow$ 1 fence post costs $\dfrac{£54}{12} = £4 \cdot 50$

$\Rightarrow$ 7 fence posts cost $7 \times £4 \cdot 50 = £31 \cdot 50$.

$\Rightarrow$ He would pay £31·50 for another seven fence posts.

Inverse Proportion

Two quantities are said to be in Inverse Proportion if as one quantity increases, the other decreases.

For example, last week in the school lottery there were 8 winners who shared the £200 prize fund. Each winner won £25. If there had been 10 winners, each would have received £20. As the number of winners increases, the winnings decrease.

Example

Three men take four days to build a low wall 48 m in length.
How long would it take two of the men to build a similar wall?

(Solution)

3 men take 4 days to build a wall
hence 1 man would take 3×4 days i.e. 12 days to build the wall

hence 2 men would take $\dfrac{12}{2}$ days i.e. 6 days to build the wall

It would take six days.

Variation

You probably met the ideas of Direct and Inverse Proportion in your earlier years at school. These topics are now extended and developed more mathematically by writing algebraic equations to describe how quantities are related. This is called **variation**, and we look first at the 'language' of variation.

Direct Variation

Direct Variation means the same as Direct Proportion. As one quantity increases, the other increases too.

Looking back to the cost (c) of a number (n) of fence posts, we would write

$$c \propto n.$$

This is read as c 'varies as' n, and means the same as c 'is directly proportional' to n. To create an equation in c and n, we write

$$c = k \times n \qquad \text{where } k \text{ is a constant.}$$

C

Inverse Variation

Inverse variation can be thought of as the opposite of direct variation, because as one quantity increases the other decreases. A simple example of inverse variation is that of the speed and time for a journey; the higher the speed the shorter the time taken.

Inverse variation is the same as inverse proportion, but written as $A \propto \dfrac{1}{B}$ and read as A varies inversely as B. Since $A \propto \dfrac{1}{B}$, then $A = \dfrac{k}{B}$, where k is a constant.

C

Joint Variation

The third category of variation is joint variation. Joint variation involves three or more quantities. For example, the time taken for a journey varies directly as the distance travelled and inversely as the speed.

This would be written as

$$t \propto \frac{d}{s}$$

$$\text{or} \quad t = \frac{kd}{s}, \text{ where } k \text{ is a constant.}$$

Example

The breaking distance, D metres, of a racing car varies directly as the square of its speed, V kilometres per hour.

The braking distance is 15 m when the speed is 50 km/h.
Calculate the braking distance when the speed is 80 km/h

(Solution)
$$D \propto V^2 \text{ or } D = k \times V^2$$
$$\Rightarrow 15 = k \times 50^2 = 2500k$$
$$\Rightarrow k = 0.006$$
Hence when $V = 80$,
$$D = 0.006 \times 80^2$$
$$= 38.4 \text{ metres}$$

Example

A train travelling on a curve exerts a horizontal force on the rails. The force, F Newtons, varies directly as the square of the train's speed, V km/h, and inversely as the radius, R metres, of the curve.

a) Write down the formula connecting F, V and R.

b) What effect does doubling the train's speed have on the horizontal force?

Example *continued* ➤

Example continued

(Solution)

a) $F \propto \dfrac{V^2}{R}$

Hence $F = \dfrac{kV^2}{R}$.

b) Now speed $= 2V$

Hence $F = \dfrac{k(2V)^2}{R} = \dfrac{k.4V^2}{R} = \dfrac{4kV^2}{R}$

Hence $F = 4\left(\dfrac{kV^2}{R}\right)$.

When compared with $F = \dfrac{kV^2}{R}$ the horizontal force is now seen to be four times greater.

Variation – Graphical Approaches

If two quantities vary directly then the graph of one against the other is a straight line through the origin.

Example In the graph shown, the volume of water leaking from a burst pipe is recorded against time.

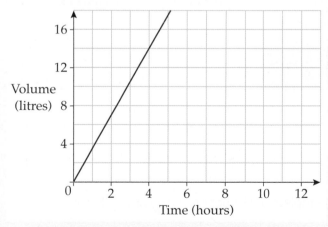

a) How much water has leaked from the burst pipe in 4 hours?

The volume of water (V litres) leaking, varies directly as the time (t hours). Hence $V \propto t$ or $V = k \times t$, where k is a constant.

b) Find the value of k and hence write down the formula connecting V and t.

Example continued ➤

Example *continued*

(Solution)

a) 14 litres (from graph).

b) Now substitute 14 for V and 4 for t in the formula $V = k \times t$ and calculate k:

thus $14 = k \times 4$

$\Rightarrow k = \dfrac{14}{4} = 3{\cdot}5$

$\Rightarrow V = 3{\cdot}5t.$

For Practice

C

If two quantities vary inversely, the graph of one against the other is more complex. (It is called a hyperbola.)

The diagram shows a gas trapped beneath a piston in a cylinder. As the piston is pushed down, to reduce the gas volume, the pressure of the gas increases. (As the piston is raised, the pressure of the gas is reduced.)

Here are some measurements of pressure and volume for a gas treated in this way.

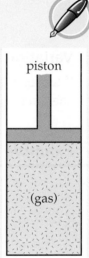

piston

(gas)

Pressure – P (kPa)	25	50	100	125	250
Volume – V	2·0	1·0	0·5	0·4	0·2

a) Draw a graph of P(y-axis) against V(x-axis).
(Do not 'join the dots'. Draw the best fitting curved line through them!)

b) (Hint: In this situation, P and V vary inversely. Hence $P \propto \dfrac{1}{V}$, or $P = \dfrac{k}{V}$.)
Use values from your graph to find the value of k.
Hence complete the equation $P = \dfrac{k}{V}$ or $P \times V = k$.

[Answer: $P \times V = 50.$]

For Practice

(The answers to the following questions are given in Appendix 2.)

5.1 The ratio of children to adults in a cinema audience on a Saturday night was $2:5$.
The total number of children and adults was 364.
How many children were in the cinema audience?

5.2 John mixes yellow paint and blue paint in the ratio $3:2$ to make green paint.
He has 23 litres of yellow paint and 14 litres of blue paint.
What is the maximum amount of green paint John can make?

5.3 A radio mast is positioned as shown on top of a tower.
A steel wire joins the top of the mast to point P on the ground.
The tower is 48 metres high.
Calculate the height of the mast, h.

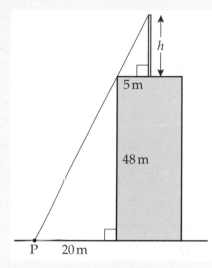

5.4 The time taken to cook a chicken in a microwave oven is inversely proportional to the power setting, P (Watts), of the oven.
It takes 8 minutes to cook a chicken in a microwave rated at 600 Watts.
How long would it take to cook a similar chicken in a microwave rated at 800 Watts?

5.5 P varies directly as X and inversely as the square root of Y.
   a) Write down the formula connecting P, X and Y.
   b) If Y is increased four times, what is the effect on P (assuming X does not change)?

For Practice continued ➤

For practice *continued*

5.6 The following graph shows the distance travelled against time, for a model car set to travel at a fixed speed.

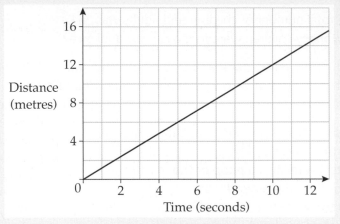

a) How far has the car travelled in 10 minutes?

The distance D (metres) travelled by the car varies directly as the time t (seconds) travelled.
Hence $D = k \times t$, where k is a constant.

b) Find k and write down the formula connecting D and t.

CALCULATION AND MEASUREMENT

In this chapter we use ideas from speed, distance, and time to introduce *rounding*, then *significant figures*, and *tolerance*.

Speed, Distance, and Time

Whenever we travel, we generally want to know how long the journey will take, how far we are travelling, and how fast we are going. So the relationships between speed, distance, and time are important to us and used by us frequently. These relationships are usually well known and, fortunately, easy to remember.

Remember

$$\text{speed} = \frac{\text{distance}}{\text{time}}$$

$$\text{distance} = \text{speed} \times \text{time}$$

$$\text{time} = \frac{\text{distance}}{\text{speed}}$$

S = speed
D = distance
T = time

Example

Irene is flying to Boston on holiday.

Her flight times are shown below.

Depart Edinburgh	Sat 24 June	2015
Arrive Paris, France	Sat 24 June	2310
Depart Paris, France	Sun 25 June	0355
Arrive Boston, USA	Sun 25 June	0545

How long will Irene have to wait in Paris before her flight to Boston?

(Solution)

Sat 24 June: time in Paris = 2310 till midnight = 50 minutes
Sun 25 June: time in Paris = midnight till 0355 = 3 hours 55 minutes

Hence total time in Paris = 50 min + 3 hours 55 min = 4 hours 45 minutes

Example

The distance between Paris and Geneva is 345 kilometres.
A train takes 3 hours and 35 minutes to travel between these cities.
Find the average speed of the train in kilometres per hour.

(Solution)

$$\text{speed} = \frac{\text{distance}}{\text{time}}$$

$$\Rightarrow \text{speed} = \frac{345}{3\cdot58} \qquad (35 \text{ minutes} = \frac{35}{60} \text{ hours} = 0\cdot58 \text{ hours})$$

$$\Rightarrow \text{speed} = 96\cdot4 \,\text{km}/\text{h}$$

Common Mistakes

Beware of using 3·35 hours for 3 hours 35 minutes (instead of 3·58 hours) in the calculation!

Rounding and Significant Figures

In the last example, the calculated value for average speed was in fact 96·368 715!
However, it has been **rounded** to **one decimal place** in the answer given, and is
showing **three significant figures**: 96·4.

**At General Level you should be able to round a decimal number to a given
number of decimal places.**

**At Credit Level you should also be able to round a number to a given
number of significant figures.**

Remember

When rounding to a number of decimal places (for example, to two decimal places),
look at the number in the **next** decimal place (third decimal place). If this number is
5 or greater, **round up**. If it is less than 5, **round down**.

Example

Round **a)** 3·684 to two decimal places   **b)** 28·065 to one decimal place.

(Solution)
a) 3·68   *Rounding to two decimal places: look at the 4; it is less than 5; so round down.*

b) 28·1   *Rounding to one decimal place: look at the 6; it is more than 5; so round up.*

Remember

When rounding to a number of significant figures (for example, rounding to three significant figures), start at the left of the number, and move right to find the first non-zero digit. This is the first significant figure. Then identify the three significant figures.

Example

◆ 62·373 has 5 significant figures – all non-zero digits are significant

◆ 90·03 has 4 significant figures – zeros between non-zero digits are significant

◆ 0·000 83 has only 2 significant figures – zeros that follow the decimal place, in numbers smaller than zero, are not significant

◆ 0·004 800 has 4 significant figures – zeros that follow the last non-zero digit at the right side of a decimal are significant

◆ 90 normally has 1 significant figure, but may have two significant figures if, at the end of a calculation, you are asked to round to 2 significant figures. In this case you need to show the second significant figure is zero.

Tolerance

In the last section, we saw that numbers can be simplified by rounding and, therefore, approximated. You should be aware that *almost all measurement* is approximate, and liable to some error. If you were asked to fill 12 bags each with 5 kg of sand, which you were to measure out, it is very unlikely that each bag would contain exactly 5 kg.

Tolerance in a measurement is the difference between the greatest and least acceptable values of the measurement.

CALCULATION AND MEASUREMENT

HOW TO PASS STANDARD GRADE MATHEMATICS

Tolerance is the amount by which a quantity is allowed to vary.

For example, a length of 25 metres measured **to the nearest metre** can be up to $\frac{1}{2}$ metre greater or smaller and can still be said to be 25 metres long.

Example

A 1·3 metre steel beam is manufactured to a tolerance of $\pm0{\cdot}05$ metres.

Within the tolerance, what is the longest length and the shortest length the beam could be?

(Solution)
Longest length = 1·3 + 0·05 = 1·35 metres.
Shortest length = 1·3 − 0·05 = 1·25 metres.

For Practice

(The answers to the following questions are given in Appendix 2.)

6.1   Jack is flying to Florida on holiday.

His flight times are shown below.

Depart Glasgow	Sat 15 July	1845
Arrive London	Sat 15 July	1930
Depart London	Sun 16 July	0815
Arrive Orlando, Florida	Sun 16 July	1640

How long will Jack have to wait in London before his flight to Florida?

6.2   The distance between Manchester and Edinburgh is 214 miles.
A coach takes 4 hours and 50 minutes to travel between these cities.
Find the average speed of the coach.

6.3   A wooden door has a tolerance of $\pm7\,\text{mm}$ in its height.
Within this tolerance what are the greatest and least heights a 1·98 metre door could be?

Chapter 7

MONEY I: FINANCIAL PROCESSES

The topic of money forms an important part of the General Level Mathematics course. Money questions are common in the General paper, and shopping, wages, hire purchase, VAT, savings and loans, foreign exchange, insurances, holidays and discount are all popular topics. Questions involving money will give you the opportunity to use basic maths operations, and to work with fractions and percentages.

At Credit Level, money does not feature quite as often. However, topics such as travel, shopping, VAT, and mobile telephone costs *have* featured in recent papers.

What You Should Know

a) for General Level: you should be familiar with
   ◆ simple interest (including calculating interest for a fraction of a year)
   ◆ wages (wage rise; commission; deductions; bonus; overtime)
   ◆ profit and loss
   ◆ insurance premiums
   ◆ currency conversions
   ◆ hire purchase, deposits and repayments
   ◆ percentage calculations including VAT and discount.

b) and in addition, for Credit Level: familiar with
   ◆ calculation of an original price, given a 'new' price and a percentage change
   ◆ compound interest.

(Interest and Currency Conversion are dealt with in the next Chapter ('Money II').)

Key Words

There is a great deal of commonly used, money-related words:
Amount, **Discount**, **Earnings**, **Wage Rise**, **Take-Home Pay**, **Profit & Loss**, **Loan** and **Borrowing**.

Other words, however, like *Per Annum (p.a.)*, *Commission*, *Surcharge* and *Depreciation/Appreciation* (Credit) are more unusual.

Additionally, students studying Credit Level Maths should also know the difference between **Simple Interest** and **Compound Interest**.

Money questions will feature in both calculator and non-calculator papers. Firstly we look at a typical calculator paper question.

Example

This year Ann's salary is £35 750.
Next year she is to get a 3·2% increase.
Calculate her salary for next year.

(Solution)
either:
$3·2\% = 0·032$
Hence increase $= 0·032 \times £35\,750$
$\qquad\qquad\qquad = £1144$
So next year's salary $= £35\,750 + £1144$
$\qquad\qquad\qquad\qquad = £36\,894.$

or:
$3·2\% = \dfrac{3·2}{100}$
Hence increase $= \dfrac{3·2}{100} \times £35\,750$
$\qquad\qquad\qquad = £1144$
So next year's salary $= £35\,750 + £1144$
$\qquad\qquad\qquad\qquad = £36\,894.$

Decide which of these two methods you prefer and stick to it.

Next, here is an example of a non-calculator paper question.

Example

Tom has a part-time job in a factory.
His basic rate of pay is £6 per hour.
At weekends he gets paid double time.
He worked 10 hours last week and his wages were £72.
How many hours did Tom work at the weekend?

(Solution)
$£72 \div 6 = £12$ *Tom was paid for 12 hours work.*
$12 - 10 = 2$ *He only worked for 10 hours, so 2 of the hours he worked he did so at double time.*
Hence Tom worked 2 hours at the weekend.
(Check) (8 hours at £6 + 2 hours at £12 = £48 + £24 = £72)

Percentage Profit

Questions on percentage profit are common.

Remember

To calculate percentage profit, you divide the profit by the cost price then multiply your answer by 100. Thus:

$$\text{percentage profit} = \frac{\text{profit}}{\text{cost price}} \times 100$$

Example

Zak bought a guitar for £340 and sold it for £385.
a) What was his profit?
b) Calculate his percentage profit.

(Solution)
a) Profit = £385 − £340 = £45.
b) Percentage Profit = $\frac{45}{340} \times 100 = 13\cdot2\%$.

Hire Purchase

When buying expensive items such as computers, furniture and cars, instead of paying cash or using a credit card you can often take out a Hire Purchase agreement with the seller. Hire Purchase is one way of paying for goods over a period of time. Usually, you pay a deposit followed by a number of instalments.

Instalments are regular payments, most commonly made monthly.

Example

A fridge/freezer may be purchased for a cash price of £386 or by a Hire Purchase agreement which involves a £50 deposit followed by 12 monthly payments of £32.

a) Calculate the total price of buying the fridge/freezer by Hire Purchase.

b) How much more than the cash price is the Hire Purchase price?

(Solution)

a) Hire Purchase price

$$= £50 + (12 \times £32)$$
$$= £50 + £384 = £434.$$

b) The extra cost for Hire Purchase = £434 − £386 = £48.

VAT

Value Added Tax (VAT) is a sales tax levied on the sale of goods and services. It is usually applied at a rate of 17·5%.

Example

a) Angela bought a digital camera which was advertised at a cost of £76·50 + VAT (17·5%).

What was the total cost of the camera? (Round you answer to the nearest penny.)

(Solution)

VAT = £76·50 × 0·175 = £13·3875   *(find VAT by multiplying £76·50 by 0·175 or*
$$\frac{17·5}{100})$$

Hence total cost £76·50 + £13·3875 = £89.8875
$$= £89·89 \text{ (to the nearest penny).}$$

Example *continued* ➤

Example *continued*

b) Angela bought her digital camera from the ALLFLASH camera shop.
She was given a price promise by the sales assistant when she bought it. The price promise read:

> PRICE PROMISE:
> If you find, within 14 days, that you can purchase the goods cheaper,
> we promise to refund the difference plus
> a further 10% of the difference.

One week later, Angela saw the same camera in the PHOTOFLASH shop priced at £85 including VAT.
How much money should be refunded to her?

(Solution)
Difference = £89·89 − £85 = £4·89
Hence refund = £4·89 + 10% of £4·89   *(difference + 10% of difference)*
         = £4·89 + £0·489 = £5·379
         = £5·38 (to the nearest penny).

Insurance Premiums

In return for the payment of an *Insurance Premium* you have the promise of money back should you suffer a loss or damage to the person or item insured. Insurance Premiums are usually paid annually, although at an extra cost, they may be paid in monthly instalments.

Home Insurance

There are two home insurance policies available. A *Buildings Policy* covers the house, garage, and any other structures on the property. A *Contents Policy* covers possessions inside the house such as furniture, appliances and clothing. Both policies cover you against a wide variety of perils, including storms, floods, lightning, fire and theft.

Example   The Scott-Sure Insurance Company advertises its premiums as follows:

> ### Scott-Sure Insurance Company
>
> Buildings Insurance Premium: 63·4p per £1000 of the value of the house insured.
> Contents Insurance Premium: £3·87 per £1000 of the value of the contents insured.

Salina's flat is valued at £95 000 and her contents at £32 000.
Calculate the total cost of insuring buildings and contents for Salina's flat.

Example *continued* ➤

Example continued

(Solution)
Buildings Insurance: $95 \times £0{\cdot}634 = £60{\cdot}23$.   *(95 (from £95 000), £0·634 (from 63·4p))*

Contents Insurance: $32 \times £3{\cdot}87 = £123{\cdot}84$.   *(32 from £32 000)*
Hence Total Cost $= £60{\cdot}23 + £123{\cdot}84 = £184{\cdot}07$.

Life Insurance

Life insurance, sometimes called life assurance, provides the payment of a sum of money after the death of the insured person. In addition, life insurance can be used as a means of investment or savings.

Car Insurance

Car insurance policies insure you against loss or damage due to fire, theft or traffic accidents. The premium depends on the value of the car and the length of time since you last made a claim for damage. A 'no claims discount' helps to keep the car insurance costs down.

Example

Gavin wants to buy insurance for his car.
The basic insurance premium will be £869, before any no claims discount.
He is entitled to four years no claims discount.
Calculate his insurance premium.

No Claims Discount

No of Years Claim Free	Discount
1	30%
2	40%
3	50%
4	60%
5 or more	65%

(Solution)
No claims discount $= 60\%$ of $£869 = £521{\cdot}40$ *(60% discount selected from table)*
Hence premium $= £869 - £521{\cdot}40 = £347{\cdot}60$.

Travel Insurance

Travel Insurance is insurance taken out by holidaymakers or travellers to cover them in case of cancellation, loss or theft of personal belongings, or medical bills.

Example

Karen and her three college friends are going back-packing around Europe for three weeks.

Use the table below to find the **total** price of their travel insurance

No of Days	Cost per person	
	Europe (£)	Worldwide (£)
1–4	13	31
5–7	17	31
8–9	17	35
10–12	21	42
13–15	23	42
Each extra period of 7 days	4	7

(Solution)

three weeks = 14 days + one extra week
Hence cost per person = £23 + £4 = £27.
Hence total cost = 4 × £27 = £108.

For Practice

(The answers to the following questions are given in Appendix 2.)

7.1 Last year Charlie's salary was £18 000.
This year he received a 4% rise.

a) How much extra money will Charlie get from his rise?

b) What is his salary this year?

7.2 Stephanie has a part-time job in an office.
Her basic rate of pay is £6·50 per hour.
For working overtime she is paid at double time.
In total she worked 15 hours last week and her wages were £146·25.
How many hours overtime did Stephanie work?

7.3 Carly bought an antique chair for £240 and sold it for £295.

a) What is her profit?

b) Calculate the percentage profit.

For Practice continued ➤

For practice continued

7.4   A large screen TV may be purchased for a cash price of £895, or by a Hire
Purchase agreement which requires a deposit of £100 and 24 payments of £39.
a)   Calculate the total cost of buying the large screen TV using the Hire
Purchase scheme.
b)   How much more than the cash price is the Hire Purchase price?

7.5   Mr and Mrs Arthur have a broken window in their house.
They receive quotes for its repair from two glaziers, Clearview Windows and
Glen Glazing.
Clearview Windows can fix it for £80 + VAT.
The Glen Glazing price is £102 including VAT, but they also offer to beat any
other 'best price' Mr and Mrs Arthur can get by *20% of the difference*
between that price and the Glen Glazing price.
a)   What is the Clearview Windows price?
b)   Mr and Mrs Arthur ask Glen Glazing to replace the window.
How much do they actually pay?

7.6   Abby and Karl insure their house with Scott-Sure Insurance Company.
Their house is valued at £115 000 and their contents at £42 000.
Calculate the total cost of insuring buildings and contents for Abby and Karl's
house.
The premiums are as follows:

Scott-Sure Insurance Company

Buildings Insurance Premium: 63·4p per £1000 of the value of the house insured.
Contents Insurance Premium: £3·87 per £1000 of the value of the contents insured.

7.7   Amanda wants to buy insurance for
her car.
The basic insurance premium will
be £729, before any no claims
discount is taken off.
She is entitled to six years no claims
discount.
Calculate her insurance premium.

No Claims Discount

No of Years Claim Free	Discount
1	30%
2	40%
3	50%
4	60%
5 or more	65%

For Practice continued ➤

For practice *continued*

7.8   Leslie and Debbie are going on holiday to Australia for four weeks.
Use the table below to find the total price of their travel insurance.

No of Days	Cost per person	
	Europe (£)	Worldwide (£)
1–4	13	31
5–7	17	31
8–9	17	35
10–12	21	42
13–15	23	42
Each extra period of 7 days	4	7

MONEY II: PERSONAL SPENDING AND SAVINGS

In this second 'Money' chapter, we look at how we deal with money on a more personal level.

Household Accounts

When you become a householder you may find that three of your most regular outgoings are for electricity, gas, and telephone.

Electricity and Gas Bills

Gas and electricity bills calculate the cost of the gas or electricity you have used over a period of time, usually about 2 or 3 months. The amount of gas or electricity is often measured in 'Units' and these Units are charged at a fixed rate. To this, VAT, currently at 5%, is added.

Example   The following chart shows Mike Appe's electricity bill.

Glen Electric Co.				
Mr M Appe 214 Old Edinburgh Road Wolfbridge			Account No. MAPP005679	
Present Reading	Previous Reading	Details of Charges		£
08690	08253	**Box a** [ ] units at 9·34p per unit		**Box b** [·]
			Standing Charge	13·56
			Sub Total	[·]
			VAT @ 5%	[·]
			Total Charge	[·]

a) Calculate the number of Units used.
b) Calculate the total charge made by the Glen Electric Co. for Mike's electricity.

Example *continued* ➤

Example *continued*

(Solution)

a) Units used = 8690 − 8253 = 437 (Box a)

b) Cost of Units = 437 × 9·34 = 4081·58p = £40·82 (Box b).
   Hence Sub Total = £40·82 + £13·56
   $$= £54·38.$$
   VAT = 5% of £54·38 = 0·05 × £54·38 = £2·72.
   Hence Total Charge = £54·38 + £2·72 = £57·10.

Common Mistakes

Do not forget to change the cost of the Units from pence into pounds sterling (£)!

Telephone Bills

Telephone bills may be received for mobile or fixed-line (house) phones. The charges for fixed-line phones depend on the time of day the call is made, whether weekday or weekend, the length of the call, and whether the call is local, national or international.

Example

Cost of calls per minute	Daytime 6am–6pm	Evenings and Night-time before 6am and after 6pm	Weekend All day Sat. and Sun.
Local Calls	3·95p	1p	1p
National Calls	7·91p	3·95p	1·5p

a) Kelly makes a local call to her friend on Monday evening at 7pm and the call lasts 2 hours 15 minutes.
   What is the cost of the call?

b) On Tuesday morning at 10am she makes a national call to her brother who works in London.
   Kelly knows it is expensive to make a daytime call and she wants to keep the cost of this call below 50p.
   What is the longest call she can make to her brother?
   Give your answer to the nearest minute.

Example *continued* ➤

Example *continued*

(Solution)
a) 2 hours 15 minutes = 135 minutes
   So cost = 135 × 1p = 135p = £1·35.

b) Cost of call per minute = 7·91p
   Maximum cost of call = 50p
   Hence length of call = 50 ÷ 7·91 = 6·32 minutes   *(divide 50p by the Daytime National rate)*

                                     = 6 minutes.   *(to the nearest minute)*

Changing Money When Travelling Abroad

Here is a table showing the currencies and conversion rates for several countries:

Country	Currency	£1 buys
Euro Zone	Euro (€)	1·42
USA	Dollar ($)	1·72
Canada	Canadian Dollars	2·06
Australia	Australian Dollars	2·27
Switzerland	Swiss Francs	2·20
Japan	Yen	191·5

Key Points

When using a table like this one
MULTIPLY to change from pounds (£) to another currency.
DIVIDE to change back to pounds (£).

Example

a) Linda is going to France on holiday.
   She changes £400 to Euros.
   Using the rate of exchange shown in the table, how many Euros will she get?

Example *continued* ➤

Example *continued*

b) While in France, Linda decides to go on a day trip to Switzerland.
She wants to change 50 Euros to Swiss Francs.
Use the exchange rate of £1 = 2·20 Swiss Francs from the table to find how many Swiss Francs Linda will get for 50 Euros.
(Round your answer to the nearest Swiss Franc.)

(Solution)

a) $400 \times 1{\cdot}42 = 568$ Euros.   (MULTIPLY by Euro rate, 1·42)

b) $50 \div 1{\cdot}42 = £35{\cdot}21$ (first change €50 into £)

$35{\cdot}21 \times 2{\cdot}20 = 77{\cdot}46$ (MULTIPLY by Swiss Franc rate, 2·20)

$= 77$ Francs.   (to nearest Swiss Franc)

Interest

Interest is money paid to you by a bank or building society for saving with them in one of their savings accounts.

Simple Interest

Interest is normally added to your account every year.
Money that you keep in an account for less than a year still gains interest, but less than you would get for a whole year.

Example

Mrs Brown has £4500 in her savings account.
The interest rate is 4·55% p.a.
How much money will Mrs Brown have in her account at the end of the year?

(Solution)

Interest = 4·55% of £4500 = $\dfrac{4{\cdot}55}{100} \times £4500 = £204{\cdot}75$.

Hence total savings = £4500 + £204·75 = £4704·75.

Example

Angela sells her flat and deposits £65 000 in her building society for seven months.
The rate of interest during this time is 3·6% p.a.
Calculate the interest she will earn in the 7 months.

(Solution)

Interest p.a. $= \dfrac{3{\cdot}6}{100} \times 65\,000 = £2340$

Hence interest for 7 months $= \dfrac{7}{12} \times 2340 = £1365$.

Remember

Candidates often do not know how to calculate the interest for a number of months. To do so you always divide by 12, the number of months in a year, then multiply by the number of months you require the interest for. To find the interest for 7 months you divide by 12 then multiply by 7, as in the example.

Compound Interest

This is interest earned on both the Principal Amount *and any interest already added*. The Principal is the original amount of savings on which interest is calculated. The total amount of money in the account grows faster this way.

Example

Mimi has £7000 invested in a savings account.
The money has been in the account for three years.
The interest rates for the three years were: Year 1 – 5·35%;
Year 2 – 4·95 %; Year 3 – 4·8%.
The savings account gathers Compound Interest.
Calculate how much Mimi will have in her account at the end of the three year period.

(Solution)

Year 1: Interest $= 5·35\%$ of £7000 $= \dfrac{5·35}{100} \times £7000 = £374·50$.

Hence amount $= £374·50 + £7000 = £7374·50$.

Year 2: Interest $= 4·95\%$ of £7374·50 $= \dfrac{4·95}{100} \times £7374·50 = £365·04$.

Hence amount $= £365·04 + £7374·50 = £7739·54$.

Year 3: Interest $= 4·8\%$ of £7739·54 $= \dfrac{4·8}{100} \times £7739·54 = £371·50$.

Hence amount $= £371·50 + £7739·54 = £8111·04$.

Mimi will have £8111·04 in her account at the end of three years.

If the interest rate is the same each year, a more efficient method can be used.

Example

Douglas invests £5000 in a savings account for 3 years.
The account has a fixed interest rate of 4·5% p.a.
Calculate the amount in Douglas's account at the end of the three year period.

Example *continued* ➤

Example *continued*

(Solution)

After one year, the account will be worth £5000 + 4·5% of £5000

$$\text{or } £5000 + \frac{4·5}{100} \times £5000$$

or £5000 + 0·045 × £5000

or £5000(1 + 0·045)

or £5000 × 1·045 (= £5225)

After the second year, the account will be worth 1·045 × £5225

or $(1·045)^2 \times £5000$.

Hence amount in savings account after 3 years = $(1·045)^3 \times 5000 = 5705·8306$

= £5705·83.

Other questions which 'compound' a percentage rate are not necessarily money questions.

Example

In a laboratory, the size of a bacteria colony increases by 4% every hour.

At the start of a trial there are 3000 bacteria.

How many bacteria will there be after 8 hours?

Round your answer to 3 significant figures.

(Solution)

Number of Bacteria = $(1·04)^8 \times 3000 = 4105·707\,151$ *(4% $\Rightarrow$ a multiplier of 1·04 each hour)*

= 4110 (rounded to 3 sig. figs.).

For Practice

(The answers to the following questions are given in Appendix 2.)

8.1   The table shows charges for a fixed-line telephone system.

Cost of calls per minute	Daytime 6am–6pm	Evenings and Night-time before 6am and after 6pm	Weekend All day Sat. and Sun.
Local Calls	3·95p	1p	1p
National Calls	7·91p	3·95p	1·5p

a)   Stuart makes a local call to his friend on Sunday afternoon and the call lasts for 1 hour and 20 minutes.
What is the cost of the call?

b)   On Friday evening he makes a national call to his mother who is in Manchester on business.
Stuart wants to keep the cost of this call below 30p.
What is the longest call he can make to his mother?
Give your answer to the nearest minute.

8.2   Here is a table showing the currencies and conversion rates for several countries:

Country	Currency	£1 buys
Euro Zone	Euro (€)	1·42
USA	Dollar ($)	1·72
Canada	Canadian Dollars	2·06
Australia	Australian Dollars	2·27
Switzerland	Swiss Francs	2·20
Japan	Yen	191·5

a)   The Mitchell family are going to Germany on holiday.
They change £650 to Euros.
Using the table, find how many Euros they will get.

b)   While in Germany, the family decides to go on a day trip to Switzerland.
They decide to change 180 Euros to Swiss Francs.
Use the exchange rate of £1 = 2·20 Swiss Francs from the table to find how many Swiss Francs they will get for their 180 Euros.
(Round your answer to the nearest Swiss Franc.)

For Practice continued ➤

For practice continued

8.3 Jacqueline has £1600 in her savings account.
The interest rate she receives is 3·75% p.a.
How much money will Jacqueline have in her account at the end of the year?

8.4 Nabila sells her car and deposits £5000 in her building society account for 10 months.
The rate of interest during this time is 4·3% p.a.
Calculate the interest she will earn in the 10 months.

8.5 Calum has £3000 invested in a cash ISA savings account.
The money has been in the account for the past four years.
The interest rates for the four years were: Year 1 – 4·8%; Year 2 – 5·15%;
Year 3 – 4·6%; Year 4 – 4·95%.
The savings account gathers Compound Interest.
Calculate how much Calum will have in his account at the end of the four year period.

8.6 A drug is administered to a patient by means of an intravenous drip. The amount of the drug in the bloodstream increases by 20% each hour.
At 8am there are 300 milligrams of the drug in the bloodstream.
How many milligrams of the drug are in the bloodstream three hours later?

ANGLES IN STRAIGHT LINE AND CIRCLE GEOMETRY

Geometry is the branch of Mathematics which deals with lines, angles, shapes and sizes of surfaces and solids.

In this chapter, we look in detail at the important part played by angles.

Key Words

An *angle* is a measure of the space between two intersecting lines. It is usually measured in degrees, close to the point of intersection of the lines.

What You Should Know

a) for General Level: you should know about
   ◆ vertically opposite, corresponding, and alternate angles
   ◆ the relationship between tangent and radius in circles
   ◆ angles in a semi-circle
   ◆ measuring the three figure bearing of one point from another
   ◆ plotting a point given its bearing and distance from another point.

b) and in addition, for Credit Level: know about
   ◆ relationships involving the centre of a circle, bisector of a chord and the perpendicular to a chord
   ◆ back bearings.

Types of Angle

Key Words

Vertically Opposite Angles (equal across a vertex); **Vertex** (point where two lines or two edges meet (a corner); **Complementary Angles** (two angles which add up to 90 degrees); **Supplementary Angles** (two angles which add up to 180 degrees and form a straight line).

Remember

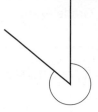

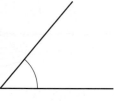

acute angle ($<90°$)

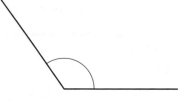

obtuse angle ($>90°$, $<180°$)

reflex angle
($>180°$, $<360°$)

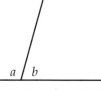

'straight' angle ($a + b = 180°$)

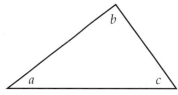

sum of angles in a triangle:
$a + b + c = 180°$

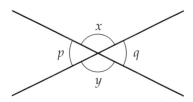

vertically opposite angles are equal
$p = q$, $x = y$

Alternate and **corresponding** angles occur when two (or more) parallel lines are cut by another line, called a *transversal*.

Corresponding angles are equal
(F – shape)

Alternate angles are equal too
(Z – shape)

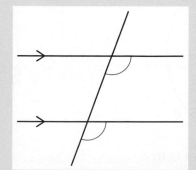

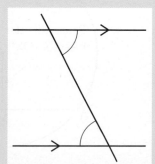

Straight Lines and Circles

There are several facts you should know about straight lines and circles which will help you to tackle these geometry questions in the Standard Grade exam.

Remember

A *tangent* to a circle is perpendicular (at right angles to) to the radius at the point of contact.

Every angle in a semi-circle is a *right* angle.

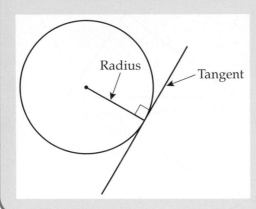

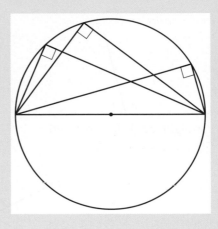

Chords and their Properties

A **chord** is a line joining two points on the circumference of a circle.
The *perpendicular bisector of a chord* passes through the centre of the circle.

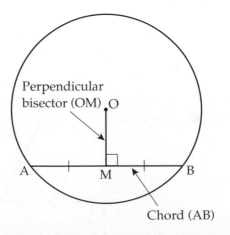

Example

A circle with centre O has AB as a diameter.

AB is 1·2 m long.

BC is a chord of the circle.

OD is parallel to AC.

Angle ABC = 28°.

Calculate the length of CE.

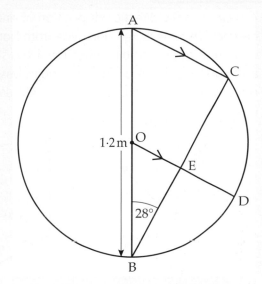

(Solution)
(firstly find the length of BC)
Angle ACB is a right-angle   (angle in a semi-circle)
Using simple trigonometry in triangle ABC:

$$\frac{BC}{1\cdot2} = \cos 28°$$

$$\Rightarrow BC = 1\cdot2 \times \cos 28° = 1\cdot06 \text{ m.}$$

(secondly)
Angle BEO = 90°   (equal to angle ACB ('F' shape))
Hence radius OD meets chord BC at 90°.
Hence CE is half of BC (OD is perpendicular bisector of BC).
Hence $CE = \frac{1}{2} \times 1\cdot06$ m $= 0\cdot53$ m.

Bearings

A *direction* can be given roughly in terms of compass points.
You should know the following directions and compass points.

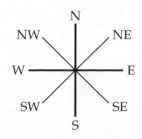

Key Words

A *bearing* is the direction or position of one point relative to another, measured in degrees, and measured clockwise from North 000° through to 360°. Bearings are written as three figure numbers, which is straightforward for angles between 100° and 360°, but more difficult for angles less than 100°. For these smaller angles one or more zero is added. For example, 45° as a bearing would be written 045° and 7° would be written 007°.

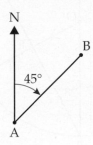

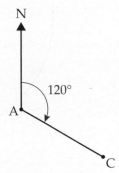

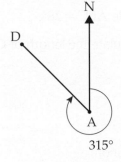

Angle NAB = 45°.
The bearing of B
from A is 045°.

Angle NAC = 120°.
The bearing of C
from A is 120°.

Angle NAD = 315°.
The bearing of D
from A is 315°.

Remember

◆ always start from North and measure in a clockwise direction
◆ always use **3** figures

Back Bearings

To illustrate *back bearings* here is a simple story.

I start at a post P and walk to a rock R.
I set off on a bearing of 045° from P.

Having sat in the sun for 10 minutes at R,
I return from rock R to post P on a bearing
of 225° (= 180° + 45°).

Since I am returning to my starting point,
the bearing of P from R (= 225°) is called
the *back bearing*.

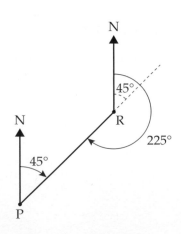

In the Standard Grade exam, bearings questions are found most often in questions involving scale drawing or trigonometry.

Hints and Tips

◆ When doing scale drawings the North line should point to the top of your page.

◆ It may be useful to continue this line slightly south beyond the point through which it is drawn. This may help you to analyse the angles in a problem more easily.

◆ When doing bearings questions you may find the use of graph paper helpful, because your 'North lines' are already in place.

ANGLES IN STRAIGHT LINE AND CIRCLE GEOMETRY

TWO-DIMENSIONAL SHAPES

Most diagrams we draw are composed of familiar two-dimensional shapes. In this chapter we deal almost entirely with two-dimensional shapes which have straight edges.

What You Should Know

a) for General Level:
   ◆ the properties of isosceles and equilateral triangles
   ◆ the properties of kite, rhombus and parallelogram
   ◆ how to recognise shapes from their nets
   ◆ how to find the area of any triangle, circle, kite, rhombus, parallelogram and composite shape (given sufficient information)
   ◆ the meanings of reflection and rotational symmetry.

b) (No additional knowledge is required for Credit Level.)

You should be very familiar with squares, rectangles and right-angled triangles. Other two dimensional (flat) shapes you might meet in a Standard Grade exam include *equilateral triangle*, *isosceles triangle*, *kite*, *rhombus*, *parallelogram*, *pentagon*, and *hexagon*.

Triangles

Most calculations with triangles involve trigonometry. Here we look at the important properties of two special types of triangle.

The Equilateral Triangle

An *Equilateral Triangle* has three sides of equal length and three equal angles, each of 60°. This may be summarised as follows, in the diagram.

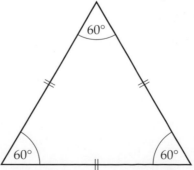

The Isosceles Triangle

An isosceles triangle has two sides of equal length and two equal acute angles as the diagrams show.

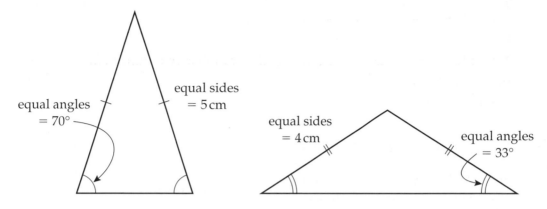

Quadrilaterals, Pentagons, Hexagons

A quadrilateral is a four-sided shape. Squares, rectangles, rhombi, kites, parallelograms, and trapeziums are all examples of quadrilaterals.

The Rhombus

A rhombus is often called a diamond shape, it looks like a square that has been 'squashed'. It has:

◆ four sides of equal length

◆ opposite sides which are parallel

◆ two pairs of equal angles.

A Rhombus has NO right angles.

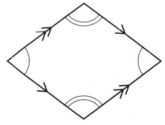

The Kite

A kite has:

◆ two pairs of sides of equal length which are next to each other

◆ one pair of equal angles.

None of its sides are parallel.

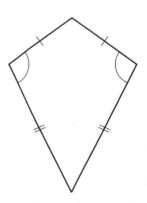

The Parallelogram

A parallelogram looks like rectangle that has been 'squashed'. It has
◆ two pairs of sides of equal length which are opposite each other
◆ opposite sides which are parallel
◆ opposite angles which are equal.

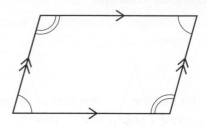

The Trapezium

A trapezium is a quadrilateral with two parallel sides.
(The trapezium shown in the diagram has two right angles (formed between the parallel sides and the shortest side). Other trapeziums have no right angles at all, however, but all have two parallel sides.)

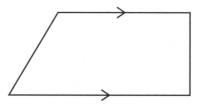

The Pentagon

A pentagon is a shape with five sides.

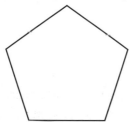

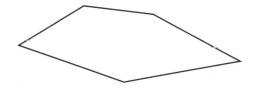

This is a regular pentagon:
◆ all five sides are equal
◆ the five internal angles are equal too ($= 108°$).

This is an irregular pentagon:
◆ all five sides are unequal
◆ all five angles are unequal too.

The Hexagon

A hexagon is a six-sided shape.

This is a regular hexagon
◆ all six sides are equal
◆ the six internal angles are equal too ($= 120°$)
◆ opposite sides are parallel.

This is an irregular hexagon
In this example:
◆ the pair of horizontal sides are equal to each other, but unequal to the other four equal sides
◆ opposite sides are still parallel.

Nets Of Shapes

If a hollow three dimensional shape is opened out and flattened, then its *Net* is produced.

You may be familiar with the nets of squares and cubes. Here we take a look at the nets of pyramids, prisms, and cylinders.

The Triangular Pyramid

This is a regular shape which has a triangle for its base and three triangles for its sides. The first diagram shows the pyramid, and the second shows how it has been opened out to reveal its net.

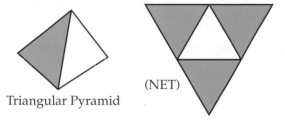

Triangular Pyramid

(NET)

The Square-based Pyramid

This is an interesting shape and is built on a square base. The Great Pyramid in Giza, Egypt, is of this design. Since it uses a square as well as triangles in its construction, it is not a regular shape.

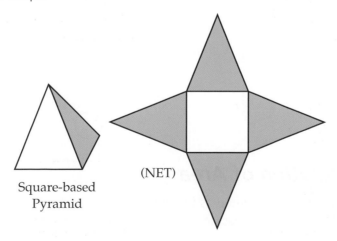

Square-based
Pyramid

(NET)

The Triangular Prism

A triangular prism is a 3D shape which has two identical parallel faces which are triangles, and a set of parallel edges connecting corresponding vertices of the two triangles.

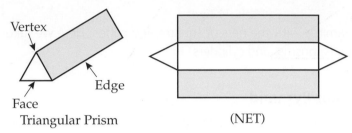

The net of a triangular prism thus consists of three rectangles of the same length, together with two congruent (identical) triangles. If the triangles are equilateral, the three rectangles are congruent (identical) too.

 Two shapes which are *congruent*, are identical in every respect of size and shape.

The Cylinder

A cylinder is a figure which has two connected circular bases which are congruent and parallel. (Examples include a soup can or a tube for shuttlecocks.)

The net of a cylinder thus consists of two congruent circles and a rectangle.

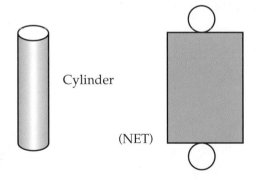

 # *The Calculation of Area*

In Mathematics area is calculated as the number of square units it takes to cover the surface of a shape. Area is measured in square centimetres (cm^2) or square metres (m^2).

The Area of a Triangle

You will recall that since a right-angled triangle is half of a rectangle, its area is calculated from the formula: $\text{Area} = \frac{1}{2} \times \text{base} \times \text{height}$ or $A = \frac{1}{2}bh$.

In fact for any triangle: $A = \frac{1}{2}bh.$

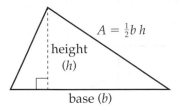

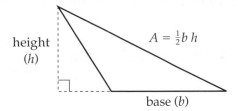

(The height is always measured at right angles to the base!)

The Area of a Parallelogram

Here Area = base × height

or $A = b\,h$

Again the height is measured at right angles to the base!

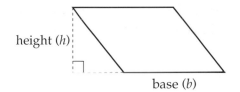

The Area of a Kite

The kite has two diagonal lines, joining opposite corners of the shape.
In this case:

Area $= \frac{1}{2} \times$ (diagonal 1 × diagonal 2)

or $A = \frac{1}{2}(d_1 \times d_2).$

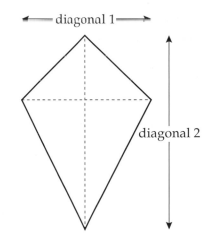

The Area of a Rhombus

The rhombus also has diagonals which join the opposite corners of the shape.

Again

Area $= \frac{1}{2} \times$ (diagonal 1 × diagonal 2)

or $A = \frac{1}{2}(d_1 \times d_2).$

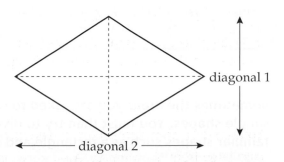

As well as being able to calculate the areas of regular shapes we have to be able to calculate the areas of shapes which are made up from more than one shape. These shapes are called **composite shapes**.

The Area of a Composite Shape

Example

Find the area of the trapezium shown in the following diagram.

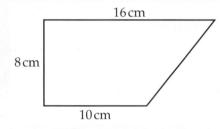

(Solution)

(The second diagram shows how the trapezium shape may be simplified.)

Hence:

$A_{(rectangle)} = l \times b$ and $A_{(triangle)} = \dfrac{1}{2}bh$

$= 10 \times 8$

$= 80 \, cm^2.$

$\quad = \dfrac{1}{2} \times 6 \times 8$

$\quad = 24 \, cm^2.$

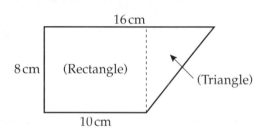

Hence total area $= 80 + 24 = 104 \, cm^2.$

Remember

Area is always measured in *square units*: for example, square centimetres (written cm^2), or square metres (written m^2).

Sometimes the areas you are asked to calculate are not those of particularly simple shapes. You must then try to divide the composite shape into more familiar shapes such as a rectangle and triangle (as above).

Example

The diagram shows part of a paved area made from congruent rhombus tiles.

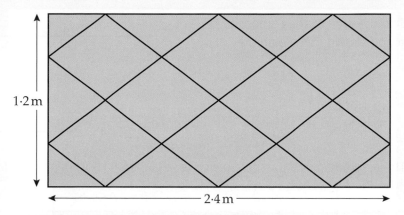

Find the area of one of the rhombus tiles.

(Solution)
(First find the lengths of the diagonals of one rhombus.
Diagonal 1 = 2·4 ÷ 3 = 0·8 m   *(3 tiles measure 2·4 m horizontally)*
Diagonal 2 = 1·2 ÷ 2 = 0·6 m   *(2 tiles measure 1·2 m vertically)*

Now $A = \frac{1}{2}(d_1 \times d_2)$.

Hence area of Rhombus tile = $\frac{1}{2} \times 0·8 \times 0·6 = 0·24 \, \text{m}^2$.

Hints and **Tips**

If you forget the formula, you can find the answer this way:
split the rhombus into two equal triangles of base 0·8 m and height 0·3 m, and use
the formula for the area of a triangle: $A = \frac{1}{2}bh$.

Hence area of $\frac{1}{2}$ rhombus $= \frac{1}{2} \times 0·8 \times 0·3 = 0·12$.

So area of rhombus $= 2 \times 0·12 = 0·24 \, \text{m}^2$.

For Practice

(The answers to the following questions are given in Appendix 2.)

10.1 Find the area of the composite shape shown in the diagram.

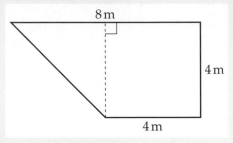

10.2 The diagram shows part of a paved area laid with congruent kite tiles.

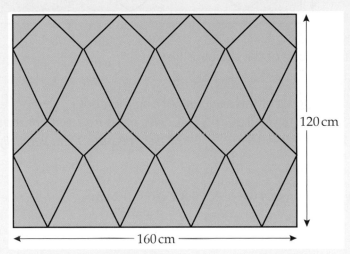

Find the area of one of the kite tiles.

Symmetry with Two-dimensional Shapes

Symmetry is the name given to an arrangement in which exactly similar shapes face each other.

There are two types of symmetry you should be familiar with for the Standard Grade exam: *reflection symmetry* and *rotation symmetry*. (Most candidates find reflection symmetry easier, but there have been several difficult reflection questions in recent years.)

For many candidates reflecting shapes in the *y*-axis seems more difficult than reflecting them in the *x*-axis. Reflecting shapes in a sloping line is the most difficult exercise of all!

Example

A is the point (1, 4), B is (5, −3) and C is (−2, −2).

a) Plot points A, B and C on the grid.

b) Reflect triangle ABC in the y-axis.

(Solution)

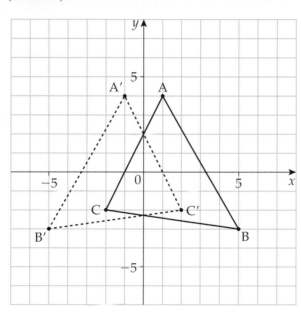

A′ is the reflection of A in the y-axis

B′ is the reflection of B in the y-axis

C′ is the reflection of C in the y-axis

Triangle A′B′C′ is the reflection of triangle ABC in the y-axis.

 Rotation symmetry may prove more difficult than reflection symmetry for many people, so a more systematic approach is recommended.

Key Words

The *order of rotational symmetry* is the number of times a shape can fit on top of its original position within a complete turn.

The *centre of symmetry* is the fixed point about which the shape is rotated.

HOW TO PASS STANDARD GRADE MATHEMATICS

Example

Part of a design is shown in the following diagram.
Complete the design so that it has rotational symmetry of order 4 about point O.

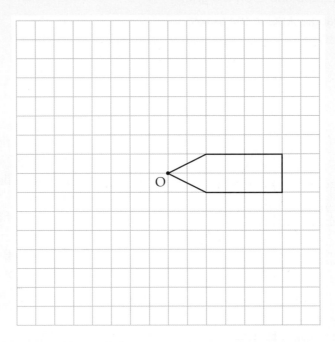

(Solution)

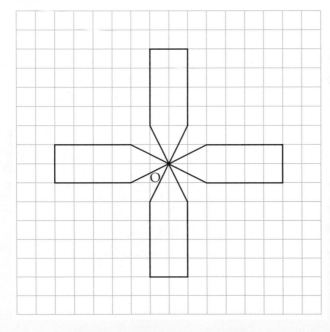

Rotation symmetry of order 4 means that the original shape will have been rotated so as to appear four times in 360°.

In other words the original shape is rotated 90°, then another 90° and then another 90°, giving a total of four identical and equally spaced shapes around the point O.

(Some people find it easier to start by rotating one line in the original shape, rather than trying to rotate the whole shape at once.)

Chapter 11

CIRCLES

The two-dimensional shapes we studied in the last chapter were straight line shapes. The circle, too, is a two-dimensional shape, and in many ways the most special shape of all.

What You Should Know

a) for General Level:
 ◆ how to calculate the circumference and area of a circle.

b) and in addition, for Credit Level:
 ◆ how to calculate arc length and sector area for part of a circle.

Revision of Circumference

The *circumference* of a circle is its perimeter. That is, it is the distance all the way round.

At General Level you will be given the formula $C = \pi d$. At Credit you must know the formula!

Example

Danny is making the cloaks for the village drama society play.

Each cloak is in the form of three-quarters of a circle.

The radius of each cloak is 125 centimetres.

The cloaks are made from purple cloth, with a strip of black edging around the perimeter.

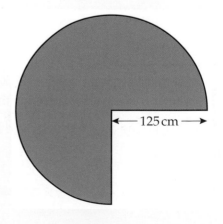

a) Calculate the length of black edging needed for each cloak.

b) The village drama society asks Danny to make 8 cloaks for the play.
 Danny has 70 m of black edging. Will he have enough for all eight cloaks?
 Give a reason for your answer.

Example continued ➤

> ### *Example* continued

(Solution)

a) radius $= 125\,\text{cm} \Rightarrow$ diameter $= 250\,\text{cm}$.

Now $C = \pi d = \pi \times 250 = 3\cdot14 \times 250 = 785\,\text{cm}$.

Hence perimeter of cloak $= \left(\dfrac{3}{4} \times 785\right) + (2 \times 125)\,\text{cm}$. *(the two straight edges are part of the perimeter)*

$$= (588\cdot75) + (250)\,\text{cm}$$
$$= 838\cdot75\,\text{cm}$$

Hence $838\cdot75\,\text{cm}$ of black edging is needed for each cloak.

b) Edging needed for 1 cloak $= 838\cdot75\,\text{cm} = 8\cdot39\,\text{m}$ *(converting to metres and rounding up)*

So edging needed for 8 cloaks $= 8 \times 8\cdot39\,\text{m} = 67\cdot12\,\text{m}$

Danny has $70\,\text{m}$ of edging so he has enough, with almost 3 metres left over.

Common Mistakes

It is easy to forget to include the two straight edges in the perimeter calculation!

Revision of Calculation of Area

Another formula given in the Formula List for General is $A = \pi r^2$, the formula for the area of a circle. At Credit you are required to know this formula too!

> ### *Example*

In Shareen's garden a circular flowerbed of radius $1\cdot8\,\text{m}$ is to be covered with tree bark.

Calculate the area to be covered by bark.

(Solution)

$A = \pi r^2$

Hence $A = \pi \times 1\cdot8^2 = \pi \times 3\cdot24 = 10\cdot2\,\text{m}^2$
$\qquad\qquad\qquad\qquad$ (to 1 dec. pl.)

Hence she needs sufficient bark to cover $10\cdot2\,\text{m}^2$.

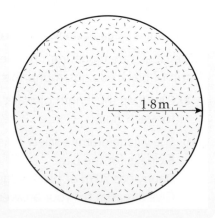

1·8 m

CIRCLES

 It is quite common to have to calculate the circumference and areas of semi-circles and quarter circles. You should check that you know how to do this.

 At Credit Level you may be required to use formulae for circumference and area in the same question.

Example

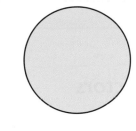

A steel band 180 cm long forms the circumference of a circular table.

←———————— 180 cm ————————→

Find the area of the circular table. Give your answer to three significant figures.

(Solution)
$C = \pi d = 180,$ *Set up Circumference formula*
so $d = \dfrac{180}{\pi} = 57.30.$ *Rearrange formula to find diameter (d)*
$d = 57.30\,\text{cm} \Rightarrow r = 28.65\,\text{cm}.$ *Find radius (r)*
Now $A = \pi r^2,$ *Use area formula*
$\Rightarrow A = \pi \times 28.65^2 = 2578.69\,\text{cm}^2$ *Substitute for r in area formula*
$\quad = 2580\,\text{cm}^2$ (to 3 sig. figs.).   *Example of rounding to 3 significant figures*

Hints and Tips

The π button on your calculator will always give you a greater degree of accuracy in the final answer than by using '3·14'. For this reason you should know where to find the π button on the calculator, and you should practise using it in questions.

HOW TO PASS STANDARD GRADE MATHEMATICS

For Practice

(The answer to the following question is given in Appendix 2.)

11.1 The circumference of a wheel is 250 cm.

Calculate the area of the wheel.

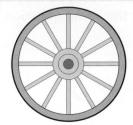

Circumference = 250 cm

Angles, Arcs and Sectors

(The remainder of this Chapter deals with Credit level topics.)

Key Words

An *arc* of a circle is a part (or fraction) of the circumference.

A *sector* of a circle is a part (or fraction) of the area of the circle, bounded by two radii and an arc (it resembles a slice of cake).

The *circle fraction* is a useful ratio in the calculation of sector angle, length of arc and areas of sector.

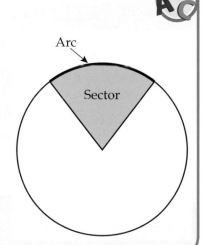

Arc

Sector

Key Points

There are three possible ways to write the circle fraction:

$$\text{circle fraction} = \frac{\text{angle within sector}}{360°} \text{ or}$$

$$\text{circle fraction} = \frac{\text{length of arc}}{\text{circumference}} \text{ or}$$

$$\text{circle fraction} = \frac{\text{area of sector}}{\text{area of circle}}$$

We now look at three important examples.
The first is on the calculation of arc length, the second on the calculation of sector area; and the third is on the calculation of the angle within the sector.
(Each example demonstrates two approaches.)

Example

a) Find the length of the arc AB.

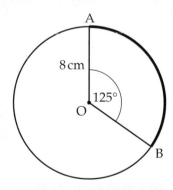

b) Find the length of the arc AB.

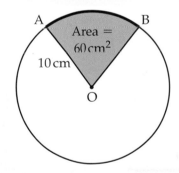

(Solution)

Circle fraction $= \dfrac{125°}{360°} = 0\cdot347.$

Circumference $= \pi d = \pi \times 16$
$= 50\cdot265.$

$\Rightarrow$ Length of arc $= 0\cdot347 \times 50\cdot265$
$= 17\cdot442$

Hence length of arc $= 17\cdot4\,\text{cm}$ (to 3 s.f.).

(Solution)

Area of circle $= \pi r^2 = \pi \times 10^2$
$= 314\cdot159.$

$\Rightarrow$ Circle fraction $= \dfrac{60}{314\cdot159}$
$= 0\cdot191.$

Now circumference $= \pi d = \pi \times 20$
$= 62\cdot83.$

Hence length of arc $= 0\cdot191 \times 62\cdot83$
$= 12\cdot0\,\text{cm}$ (to 3 s.f.).

Example

a) Find the area of sector AOB.

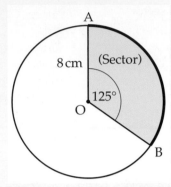

b) Find the area of sector AOB.

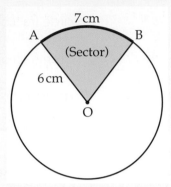

(Solution)

Circle fraction $= \dfrac{125°}{360°} = 0{\cdot}347.$

Area of circle $= \pi r^2 = \pi \times 8^2$

$= 201{\cdot}06.$

Hence area of sector $= 0{\cdot}347 \times 201{\cdot}06$

$- 69{\cdot}768$

$= 69{\cdot}8\,\text{cm}^2$

(to 3 s.f.).

(Solution)

Circumference $= \pi d = \pi \times 12$

$= 37{\cdot}699.$

Hence circle fraction $= \dfrac{7}{37{\cdot}699}$

$= 0{\cdot}186.$

Area of circle $= \pi r^2 = \pi \times 6^2$

$= 113{\cdot}097.$

Hence area of sector $= 0{\cdot}186 \times 113{\cdot}097$

$= 21{\cdot}0\,\text{cm}^2$

(to 3 s.f.).

Example

a) Find the size of angle AOB.

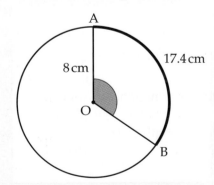

b) Find the size of angle AOB.

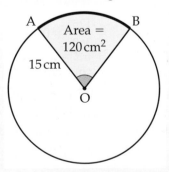

Example continued ➣

Example *continued*

(Solution)
Circumference $\quad = \pi d = \pi \times 16$
$\qquad\qquad\qquad = 50{\cdot}265.$

Hence circle fraction $= \dfrac{17{\cdot}4}{50{\cdot}265}$
$\qquad\qquad\qquad = 0{\cdot}346.$

Hence angle AOB $\quad = 0{\cdot}346 \times 360°$
$\qquad\qquad\qquad = 124{\cdot}6°$
$\qquad\qquad\qquad = 125°$ (to 3 s.f.).

(Solution)
Area of circle $\qquad = \pi r^2 = \pi \times 15^2$
$\qquad\qquad\qquad = 706{\cdot}858.$

Hence circle fraction $= \dfrac{120}{706{\cdot}858}$
$\qquad\qquad\qquad = 0{\cdot}1698.$

Hence angle AOB $\quad = 0{\cdot}1698 \times 360°$
$\qquad\qquad\qquad = 61{\cdot}128°$
$\qquad\qquad\qquad = 61{\cdot}1°$ (to 3 s.f.).

Most circle questions you will meet in the Credit exam will be set in a practical context, and will often be the reasoning (problem-solving) type.

Example

The diagram shows the cross-section of a spotlight bulb.

Find the length of the arc AB.

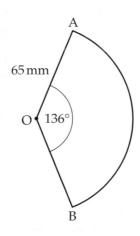

(Solution)
Radius $= 65\,\text{cm} \Rightarrow$ diameter $= 130\,\text{cm}.$

Circle fraction $= \dfrac{\text{Angle within sector}}{360°} = \dfrac{136°}{360°} = 0{\cdot}38.$

Circumference of circle $= \pi d = \pi \times 130 = 408{\cdot}4\,\text{cm}$

Hence length of arc AB $= 0{\cdot}38 \times 408{\cdot}4 = 155{\cdot}19\,\text{cm}$
$\qquad\qquad\qquad\qquad = 155\,\text{cm}$ (to 3 s.f.).

Example

The tip of a clock pendulum travels along an arc of a circle with centre O.

The length of the pendulum, OA, is 40 centimetres.

The pendulum swings from position OA to position OB.
The length of the arc traced out by the tip of the pendulum is 14 centimetres.
Find the angle through which the pendulum swings.

Example *continued* ➤

Example *continued*

(Solution)

Radius = 40 cm $\Rightarrow$ diameter = 80 cm.

$C = \pi d = \pi \times 80 = 251 \cdot 3$ cm.

Circle fraction $= \dfrac{\text{arc length}}{\text{circumference}} = \dfrac{14}{251 \cdot 3} = 0 \cdot 056$.

Hence angle of swing (= circle fraction $\times$ 360°) = $0 \cdot 056 \times 360° = 20 \cdot 2°$ (to 3 s.f.).

Example

A new wheel is being manufactured for a steam train.

The wheel (diameter 1·5 m) is to have a segment of solid steel welded to its inside edge as shown in the following diagram. (The steel segment has been shaded.)

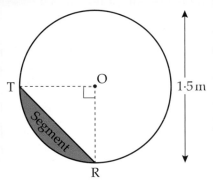

The angle subtended at the centre of the wheel (O) by the segment is 90°.

Find the area of the segment.

(Solution)

Diameter = 1·5 m $\Rightarrow$ radius = 0·75 m.

Circle fraction $= \dfrac{90°}{360°} = \dfrac{1}{4} = 0 \cdot 25$.

Area of **sector** OTR = $0 \cdot 25 \times \pi r^2 = 0 \cdot 25 \times \pi \times 0 \cdot 75^2 = 0 \cdot 442$ m^2.

Area of **triangle** OTR $= \dfrac{1}{2}bh = \dfrac{1}{2} \times 0 \cdot 75 \times 0 \cdot 75 = 0 \cdot 281$ m^2.   (OT = OR = 0·75 m)
(radii)

Hence area of segment = $0 \cdot 442 - 0 \cdot 281 = 0 \cdot 161$ m^2.   (sector area – triangle area)

For Practice

(The answers to the following questions are given in Appendix 2.)

11.2 The diagram shows a design for a patio.

Find the length of the arc PQ.

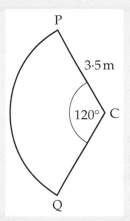

11.3 The tip of the hour hand of a clock travels along an arc of a circle with centre O.

The length of the hour hand, OC, is 25 centimetres.
The hour hand moves from position OC to position OD.
The length of the arc travelled by the tip of the hour hand is 18 centimetres.
Find the angle through which the hour hand turns.

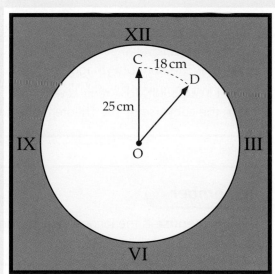

11.4 A metal disk is being manufactured. The shaded segment shown in the diagram is to be removed.
Find the area of this segment.

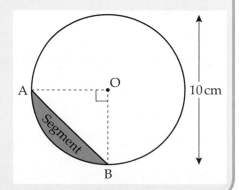

THEOREM OF PYTHAGORAS

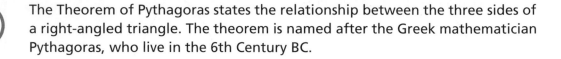

i The Theorem of Pythagoras states the relationship between the three sides of a right-angled triangle. The theorem is named after the Greek mathematician Pythagoras, who live in the 6th Century BC.

! **This theorem may be useful to you in answering questions involving two-dimensional shapes. It will also be very helpful in the later work on Trigonometry.**

What You Should Know

a) for General Level:
   ◆ The Theorem of Pythagoras.
b) and in addition, for Credit Level:
   ◆ the converse of The Theorem of Pythagoras.

Remember

The **hypotenuse** is the longest side of a right-angled triangle; *it is always opposite the right angle*.

Key Points

The Theorem of Pythagoras states that:
in a right-angled triangle $c^2 = a^2 + b^2$.

You can be asked to calculate the length of either the hypotenuse or one of the shorter sides in a right-angled triangle.

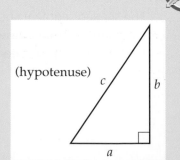

To use the Theorem of Pythagoras you need to know the lengths of two sides of a right-angled triangle. Then you can calculate the length of the third side.

Example

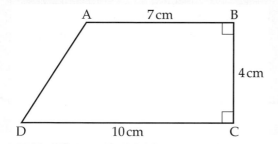

ABCD is a trapezium in which
◆ AB = 7 centimetres
◆ BC = 4 centimetres
◆ CD = 10 centimetres
◆ angles ABC and BCD are both right angles.

Calculate the length of side AD.
(Do not use a scale drawing.)

(Solution)

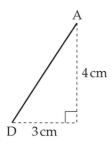

(Draw a vertical construction line from A to DC so as to form a right-angled triangle with AD as the hypotenuse.)

then $AD^2 = 4^2 + 3^2$ *(using the Theorem of Pythagoras)*

$\Rightarrow AD^2 = 25$

$\Rightarrow AD = \sqrt{25}$

$\Rightarrow AD = 5\,\text{cm}.$

Remember

Remember that the symbol $\sqrt{\ }$ means take the square root. You should know the square root of (at least) the following numbers: 1, 4, 9, 16, 25, 36, 49, 64, 81, 100. (*Do* you know them?)

Common Mistakes

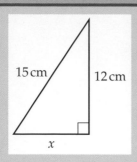

15 cm 12 cm

x

You must **never** write

$x^2 = 15^2 + 12^2$!!

(In this case, $15^2 = 12^2 + x^2$, leading to $x^2 = 15^2 - 12^2$)

Example

a) On the grid below, plot the points A(-4, -3) and B(3, 2).

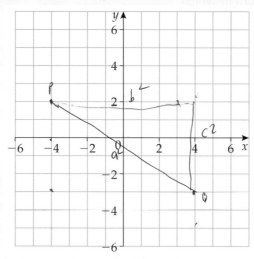

b) Calculate the length of the line AB.

(Solution)

a) A and B are plotted as shown.

b) Now draw in lines AC and BC, thus forming a right-angled triangle ABC.
By inspection AC = 7
   BC = 5.
Hence $AB^2 = 7^2 + 5^2$
   $= 49 + 25$
   $= 74$.
Hence $AB = \sqrt{74} = 8{\cdot}6023$
   $= 8{\cdot}6$ (to 1 d.p.).

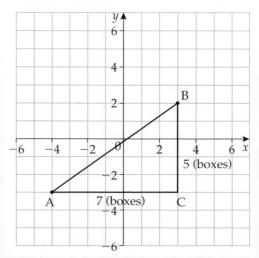

Example

The diagram shows an arched window which an architect has designed for a new building.

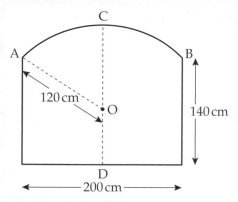

AB is an arc of a circle with centre O.
The width of the window is 200 centimetres, the height of its vertical edge is 140 centimetres and the radius OA is 120 centimetres.
Calculate the height CD of the window.

(Solution)
First we add the dotted construction line AE, at 90° to CD.

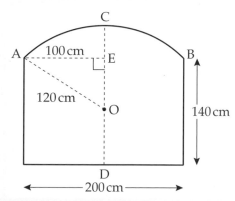

$AE = \frac{1}{2} \times 200 = 100$ cm.
By Pythagoras, $120^2 = 100^2 + EO^2$
$\Rightarrow EO^2 = 120^2 - 100^2$
$\qquad = 4400.$
So $EO = \sqrt{4400} = 66 \cdot 3$ cm.
Now $EO + OD = 140$ cm (the vertical edge).
So $OD = 140 - 66 \cdot 3 = 73 \cdot 7$ cm.
Also $CD = OD + $ (radius) OC.
Hence $CD = 73 \cdot 7 + 120 = 193 \cdot 7$ cm.

C The Converse of the Theorem of Pythagoras

At Credit Level, as well as being able to use the Theorem of Pythagoras to calculate the lengths of sides of triangles, you also have to be able to show why some triangles are right-angled.

Key Points

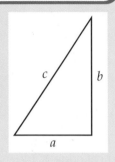

If for a triangle $c^2 = a^2 + b^2$,
then the triangle is right-angled (and the right angle is the angle opposite side c).

This is the converse of the Theorem of Pythagoras.

Example

The dimensions of a triangular sheet of metal are shown in the diagram.

Is this sheet of metal a right-angled triangle?

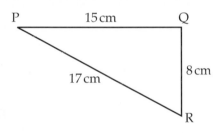

(Solution)
Here $PQ^2 + QR^2 = 15^2 + 8^2$
$$= 225 + 64$$
$$= 289$$
$$= 17^2.$$
Hence $PQ^2 + QR^2 = PR^2$.

Hence triangle PQR is right-angled at Q.

For Practice

(The answers to these questions are given in Appendix 2.)

12.1 VWXY is a parallelogram:
- ◆ VZ = 7 metres
- ◆ YZ = 3.5 metres
- ◆ angle VZY is 90°.

Calculate the length of side VY.
(Do not use a scale drawing.)

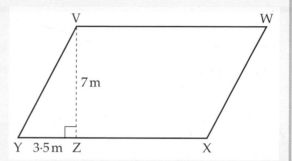

12.2 The diagram shows the entrance to a tunnel.
PQ is an arc of a circle with centre O.
The width of the tunnel is 2 metres, the height of the vertical edge is 1·5 metres and the radius OP is 1·3 metres.
Calculate the height RS of the tunnel.

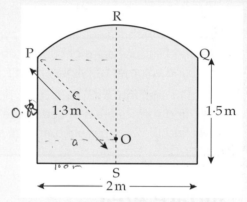

12.3 a) Using a piece of squared paper plot the points P(−5, 2) and Q(4, −3).

   b) Calculate the length of the line PQ.

12.4 The dimensions of a triangle are shown in the diagram.

   Is this triangle right-angled?

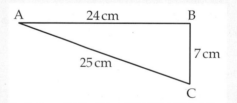

THREE-DIMENSIONAL SHAPES

The last three chapters were concerned with two-dimensional shapes. Some of the ideas which we met in these chapters are now applied to the more complex three-dimensional shapes.

What You Should Know

a) for General Level: how to
   - ◆ work with triangular prisms, pyramids, cylinders, and more complex shapes
   - ◆ find surface areas of cubes, cuboids, cylinders and triangular prisms
   - ◆ find volumes of cylinders and triangular prisms.

b) and in addition, for Credit Level: how to
   - ◆ find surface areas of composite solids
   - ◆ find the volumes of composite solids.

Surface Area

When calculating the area of a three-dimensional shape, we first check how many identical faces there are.

For example, the opposite faces in a cuboid are identical.

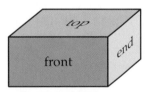

- ◆ the front and back are congruent (same size and shape)
- ◆ the top and bottom are congruent
- ◆ both ends are congruent.

So cuboids have **3 pairs** of congruent faces.

(This is the kind of useful information we use in calculating the surface area of a three-dimensional shape.)

Example

A large advertising sign is in the shape of a triangular prism.

Alex has to paint the sign with white gloss paint before the advert transfer is applied.

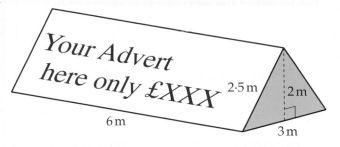

He will paint the rectangular faces and the triangular ends but not the bottom.

a) Calculate the total area he will paint with white gloss.

b) Each tin of paint covers $10\,m^2$.
   What is the minimum number of tins of paint Alex will need?

(Solution)

a) $A_{\text{rectangle}} = 6 \times 2{\cdot}5 = 15\,m^2$.   ($A = lb$)

$A_{\text{triangle}} = \dfrac{1}{2} \times 3 \times 2 = 3\,m^2$.   ($A = \dfrac{1}{2}bh$)

Hence total area $= (2 \times 15) + (2 \times 3) = 36\,m^2$.   *(two rectangles and two triangles)*

b) Alex will need 4 tins.   *(3 tins will only cover $30\,m^2$)*

The Surface Area of a Cylinder

The surface area of a cylinder is a little more complicated to calculate than the surface area of the triangular prism in the last Example.

(Here we have a closed cylinder which has a top and bottom, rather than just a hollow tube.)

The top and bottom of a cylinder are both circles and their areas are easy to calculate.

Each has area πr^2, so the *total area for top and bottom is $2\pi r^2$*.

The area of the curved surface of the cylinder needs a little more thought however.

When you cut open the **side** of the cylinder and flatten it out you get a rectangle.

The height of the rectangle is the same as the height of the cylinder ($= h$). (See diagram.)

The width of the rectangle is just the circumference of either of the circles. Thus width $= C = \pi d$.

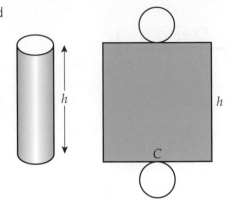

Hence *curved surface area of cylinder* $=$ width $\times$ height (of rectangle) $= \pi d \times h$.

Hence the total surface area of the cylinder ($=$ top $+$ bottom $+$ curved surface)
$$= 2\pi r^2 + \pi dh.$$

Example

A closed metal tube is in the shape of a cylinder.

It is 50 centimetres high and 15 centimetres in diameter.

Find the total surface area of the metal tube.

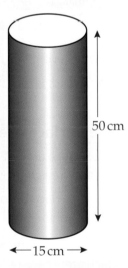

(Solution)
Diameter (d) $= 15\,\text{cm} \Rightarrow$ radius (r) $= 7{\cdot}5\,\text{cm}$.

Area of top $= \pi r^2 = \pi \times 7{\cdot}5^2 = 176{\cdot}7$.

Area of curved surface $= \pi dh = \pi \times 15 \times 50 = 2356{\cdot}2$.

Hence total surface area $= (2 \times 176{\cdot}7) + 2356{\cdot}2 = 2709{\cdot}6\,\text{cm}^2$.

Volume

The volume of a shape is the amount of space it occupies. We usually measure this in cubic centimetres (cm^3) or cubic metres (m^3).

You will already be familiar with the use of the volume formulae for cubes and cuboids from your earlier secondary school maths courses.

Remember

Volume of a cuboid = length × breadth
$$\qquad\qquad\quad × \text{height}$$
$$= l × b × h$$
or $V = l\,b\,h$.

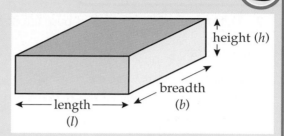

Volume of a cube = length × length × length
$$= l × l × l$$
or $V = l^3$.

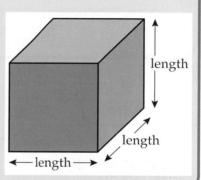

Cylinders and triangular prisms are the two other three-dimensional shapes whose volumes you may frequently be asked to calculate.

Key Points

1 Volume of a Cylinder = area of base × height
$$= \pi r^2 × h$$
or $V = \pi r^2 h$.

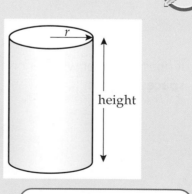

Key Points continued ➤

HOW TO PASS STANDARD GRADE MATHEMATICS

Key Points *continued*

2 Volume of a Prism = area of base (cross-section) × height
$$= A \times h$$
$$\text{or } V = Ah.$$

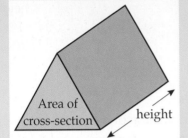

Area of cross-section   height

Example

A tube for potato crisps is in the shape of a cylinder.
It is 30 centimetres high and its diameter is 7 centimetres.
Calculate the volume of the tube in cubic centimetres.
(Round your answer to the nearest cubic centimetre.)

(Solution)
$$V = \pi r^2 h$$
$$= \pi \times 3 \cdot 5^2 \times 30$$
$$= 1154 \cdot 53$$
$$= 1155 \text{ cm}^3 \text{ (to the nearest cubic centimetre)}.$$

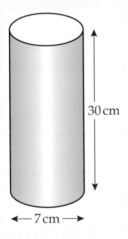

30 cm

←— 7 cm —→

It doesn't matter whether or not the tube has a top. Volume is a measure of the
space occupied.

Example

A jar is in the shape of a cylinder with diameter 8 centimetres and height 13 centimetres.

a) Calculate the volume of the jar.

b) 500 millilitres of water are poured into the jar. Calculate the depth of water in the jar.

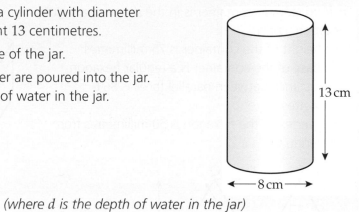

13 cm

8 cm

(Solution)

a) $V = \pi r^2 h$

$= \pi \times 4^2 \times 13$

$= 653 \cdot 45 \, \text{cm}^3$.

b) $500 = \pi \times 4^2 \times d$ (where d is the depth of water in the jar)

$\Rightarrow d = \dfrac{500}{\pi \times 16}$

$\Rightarrow d = 9 \cdot 95 \, \text{cm}$.

The depth of water is therefore $9 \cdot 95 \, \text{cm}$.

Example

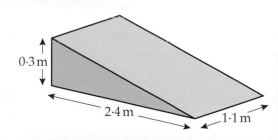

0·3 m

2·4 m

1·1 m

Darren is constructing a concrete ramp outside his house.

The ramp is in the shape of a triangular prism.

Calculate the volume of concrete required for the ramp.

(Solution)

Area of cross-section $= \dfrac{1}{2} b h = \dfrac{1}{2} \times 2 \cdot 4 \times 0 \cdot 3 = 0 \cdot 36 \, \text{m}^2$.

Volume of prism: $V = A h$

$= 0 \cdot 36 \times 1 \cdot 1$

$= 0 \cdot 396 \, \text{m}^3$

The volume of concrete required is therefore $0 \cdot 396 \, \text{m}^3$

THREE-DIMENSIONAL SHAPES

Example

C

A food storage container is in the shape of a regular hexagonal prism.
The height of the container is 75 millimetres.
The base of the container is a regular hexagon.
The distance between parallel faces is 86·6 millimetres.
Each vertex of the hexagon is 50 millimetres from the centre O.

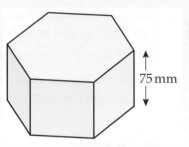

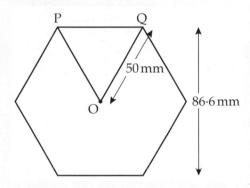

a) Calculate the volume of the container.
   (Give your answer rounded to the nearest cubic centimetre.)

b) A volume of 390 millilitres of liquid is now poured into the container.
   Calculate the depth of liquid in the container to the nearest millimetre.

(Solution)

a) A regular hexagon consists of six equilateral triangles.
   So triangle OPQ is equilateral.
   Hence PQ = 50 mm, and the height of triangle OPQ = 43·3 mm ($\frac{1}{2}$ of 86·6 mm).
   Hence area of triangle OPQ ($= \frac{1}{2}bh$) $= \frac{1}{2} \times 50 \times 43·3 = 1082·5$ mm².
   Area of hexagon = 6 × area of triangle OPQ = 6 × 1082·5 = 6495 mm².
   Hence volume of prism ($= Ah$) = 6495 × 75 = 487 125 mm³.
   Hence volume = 487 cm³. (*1 cm³ = 1000 mm³*)

b) 390 millilitres = 390 cm³ = 390 000 mm³ (*1 cm³ = 1000 mm³*)
   Using $V = Ah$ for the volume of liquid contained,
   then 390 000 = 6495 × d (where d is the depth of liquid).
   Hence d = 390 000 ÷ 6495 = 60·05 mm.
   The depth of liquid in the container is thus 60 mm (to the nearest millimetre).

For Practice

(The answers to these questions are given in Appendix 2.)

13.1 A biscuit tin is in the shape of a cylinder.

It is 20 centimetres high and 7 centimetres in diameter.

Find the curved surface area of the biscuit tin.

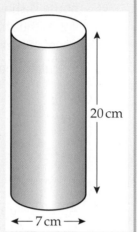

20 cm

7 cm

13.2 A spice container is in the shape of a cylinder.
It is 8 centimetres high and its diameter is
4 centimetres.
Calculate the volume of the container in
cubic centimetres.

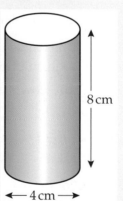

8 cm

4 cm

13.3 A supermarket sells some cheese
in wedges as shown in the
diagram.

The wedge of cheese is in the
shape of a triangular prism.

Calculate the volume of the
wedge of cheese.

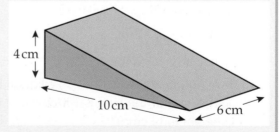

4 cm

10 cm   6 cm

For Practice continued ➤

For practice *continued*

13.4 The archway between two rooms in Sarah's house is shown in the diagram.
It is in the shape of a rectangle with a semi-circle above it.

Calculate the perimeter of the archway.

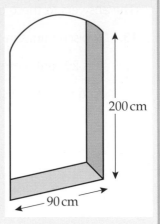

200 cm

90 cm

13.5 A traffic sign is in the shape of a triangular prism, with two triangular ends.

a) Calculate the total surface area of the sign.

The whole sign is to be painted.
One tin of paint covers 2500 cm².

b) How many tins of paint are needed to paint the sign?

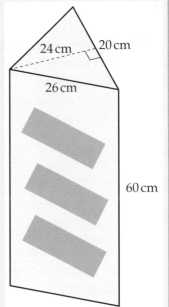

24 cm 20 cm

26 cm

60 cm

13.6 A storage container has an elliptical base. The area of the base is 100 cm² and the height is 14 cm.

a) Calculate the volume of the container.

b) 500 millilitres of orange juice is poured in.
Calculate the depth of orange juice in the container.

Orange Juice

14 cm

Chapter 14

SOLVING EQUATIONS

Algebra is the branch of mathematics in which we use letters to represent unknown quantities. We then set up and solve equations to find the value of any particular unknown quantity. Increasing our understanding of algebra will help us in analysing and solving problems (the Reasoning and Enquiry (RE) element in the exam paper). The equations we deal with here are either **linear** (such as $2x + 7 = 13$) or **quadratic** (such as $x^2 - 6x + 5 = 0$).

What You Should Know

a) for General Level: how to
  ◆ collect like terms
  ◆ multiply expressions
  ◆ factorise using a common factor
  ◆ solve simple equations and inequalities.

b) and in addition for Credit Level: how to
  ◆ factorise a difference of two squares and a trinomial
  ◆ solve simultaneous equations and quadratic equations.

Key Words

Terms ★ Brackets ★ Factorise ★ Common Factor ★ Difference of Squares ★ Trinomial ★ Solve ★ Equation ★ Inequality ★ Quadratic ★ Simultaneous ★ Graphically ★ Algebraically ★ Variable.

Before beginning work on equations we look at some important processes which will be used.

Collecting Like Terms

$4x$, $6x$ and $-9x$ are called LIKE terms.
Only LIKE terms can be *added* or *subtracted*.

Example

(unlike terms)

(i) $4x + 6x$

$= 10x$

(ii) $8a - 9a$

$= -a$

(iii) $2x + 3a$

$= 2x + 3a.$

Multiplication of Terms

ANY terms, like or unlike, can be multiplied together.

Example

(i) $a \times b$

$= ab$

(ii) $t \times t$

$= t^2$

(iii) $4 \times 2y$

$= 8y$

(iv) $2c \times 5d$

$= 10cd.$

Brackets

Sometimes you may need to remove brackets from an expression. To do this simply multiply each term in the bracket by the *number* outside it.

Example

(i) $3(a + 5)$

$= 3a + 15$

(ii) $6(x - 2y)$

$= 6x - 12y$

(iii) $4 + 5(m - 3)$

$- 4 + 5m - 15$

$= 5m - 11.$

For more difficult work involving brackets, you may have to multiply each term in the bracket by the *term* outside it.

Example

(i) $-x\,(x + 4y)$

$= -x^2 - 4xy$

(ii) $(a + b)(x + y)$

$= a(x + y) + b(x + y)$

$= ax + ay + bx + by$

(iii) $(x + 3)(x - 2)$

$= x(x - 2) + 3(x - 2)$

$= x^2 - 2x + 3x - 6$

$= x^2 + x - 6$

(iv) $(c + 3)\,(c^2 - 5c + 2)$

$= c(c^2 - 5c + 2) + 3\,(c^2 - 5c + 2)$

$= c^3 - 5c^2 + 2c + 3c^2 - 15c + 6$ (*now collect LIKE terms*)

$= c^3 - 2c^2 - 13c + 6.$

Factorisation

When we multiply out brackets, we are expanding an expression.
Sometimes, however, we wish to 'tidy' an expression of terms by doing the reverse.
This is called *factorisation*.

Factorising Using a COMMON FACTOR

Example

(i) $4a + 4b$
      $= 4(a + b)$

(ii) $6x - 15$
      $= 3(2x - 5)$

(iii) $10p + 5pr$
      $= 5p(2 + r)$

 Observe how *all* the common factors are taken outside the brackets so that the expressions are factorised fully.

Example

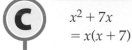

$x^2 + 7x$
$= x(x + 7)$

 ### Factorising DIFFERENCES OF SQUARES and TRINOMIALS

 As in so many other areas of your mathematics course, the more you practise, the easier it becomes. The following worked examples should serve as useful reminders.

Example

(i) $a^2 - b^2$
      $= (a + b)(a - b)$

(ii) $4p^2 - 9q^2$
      $= (2p + 3q)(2p - 3q)$

(iii) $x^2 + 3x - 10$
      $= (x - 2)(x + 5)$

(iv) $2x^2 + x - 3$
      $= (2x + 3)(x - 1)$.

Solving Simple Linear Equations and Inequalities

An **equation** can be thought of quite simply as a set of balanced scales.

The scales remain balanced provided we add or subtract the same weight to or from each side. Similarly, an equation remains an equation provided we do the same to each side.

Example

Solve for x:

(i) $2x + 7 = 13$
$\Rightarrow 2x = 6$
$\Rightarrow x = 3$

(ii) $4x - 6 = x + 9$
$\Rightarrow 3x - 6 = 9$
$\Rightarrow 3x = 15$
$\Rightarrow x = 5.$

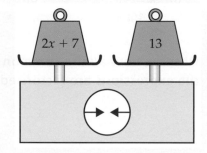

Example

$$7x - 3(x - 4) = 10$$
$$\Rightarrow 7x - 3x + 12 = 10$$
$$\Rightarrow 4x + 12 = 10$$
$$\Rightarrow 4x = -2$$
$$\Rightarrow x = -\frac{2}{4}$$
$$x = -\frac{1}{2}.$$

An **inequality** can be thought of quite simply as a set of *unbalanced* scales. The scales remain unbalanced *in the same way* provided we add or subtract the same weight to or from each side. Similarly, an inequality remains an inequality provided we do the same to each side.

Example

$$2x + 5 < 8$$
$$\Rightarrow 2x < 3$$
$$\Rightarrow x < \frac{3}{2}.$$

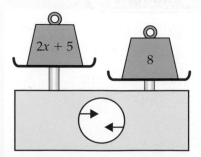

HOW TO PASS STANDARD GRADE MATHEMATICS

Example

$$x + 3 \geqslant 5x - 2$$
$$\Rightarrow \qquad 3 \geqslant 4x - 2 \qquad \text{(gather 'x' terms on the side with the greater number)}$$
$$\Rightarrow \qquad 5 \geqslant 4x$$
$$\Rightarrow \qquad \frac{5}{4} \geqslant x$$
$$\text{or} \qquad x \leqslant \frac{5}{4}.$$

Remember

$x > 7$ means that x **is greater than** 7
$y < 4$ means that y **is less than** 4
$x \geqslant 7$ means that x **is greater than or equal to** 7
$y \leqslant 4$ means that y **is less than or equal to** 4.

Solving Simultaneous Equations

Some problems introduce two equations in two unknowns which have to be solved **simultaneously**, or together.

These are called *simultaneous equations*.

Simultaneous equations can be solved by either of two methods: graphically or algebraically. We now look at these.

Solving Simultaneous Equations Graphically

We begin with a very simple example which reminds you of the equations of lines parallel to axes.

Example

Solve the simultaneous equations $x = 1$ and $y = 3$.

(Solution)
The graph of each equation is drawn.
It shows that the lines $x = 1$ and $y = 3$ intersect at (1, 3).

The solution of the simultaneous equations $x = 1$ and $y = 3$
is thus (1, 3).

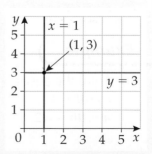

Example

Solve the simultaneous equations $y = x + 1$ and $y = -2x + 7$.

(Solution)
To draw these lines, we first calculate the y-values corresponding to simple x-values for each line. These are shown in the tables.

x	0	1	2	3
$y = x + 1$	1	2	3	4

x	0	1	2	3
$y = -2x + 7$	7	5	3	1

(Now plot (0, 1), (1, 2), (2, 3), (3, 4).

(Now plot (0, 7), (1, 5), (2, 3), (3, 1).

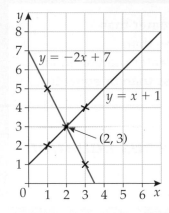

From the graph, we see that the solution of the simultaneous equations is (2, 3).

Solving Simultaneous Equations Algebraically

This approach is more commonly used. The method involves equalising either the 'x' terms or 'y' terms in each equation, and then adding or subtracting the equations to eliminate the equalised terms. The following examples should be studied carefully.

Example

(i) Solve the equations for a and b:

$a + 4b = 14$

$a - 4b = -2.$

(ii) Solve the equations for x and y:

$5x + 3y = 12$

$7x - 4y = 25.$

Example continued ⟩

Example continued

(Solution)

(i) The terms here are already equalised. Adding the equations gives

$$2a = 12$$
$$\Rightarrow \quad a = 6.$$

Substitute this value in the first equation.

Hence $6 + 4b = 14$

$$\Rightarrow \quad 4b = 8$$
$$\Rightarrow \quad b = 2.$$

The solution is therefore $a = 6$, $b = 2$.

(ii) $5x + 3y = 12$ (1)
   $7x - 4y = 25$ (2)

Multiplying (1) by 4 and (2) by 3:

$20x + 12y = 48$ (3)
$21x - 12y = 75$ (4)

Adding (3) and (4) gives:

$$41x = 123$$
$$\Rightarrow \quad x = \frac{123}{41} = 3.$$

Substitute this value into (1):

hence $5 \times 3 + 3y = 12$

$$\Rightarrow \quad 15 + 3y = 12$$
$$\Rightarrow \quad 3y = -3$$
$$\Rightarrow \quad y = -1.$$

The solution is therefore $x = 3$, $y = -1$.

C Solving Quadratic Equations

Key Words and Definitions

A **quadratic equation** is an equation in which the highest power of the variable is 2. Although we mostly use x as the variable, you may come across a quadratic equation which uses another variable. You may find an equation with t^2 or p^2.

The *Standard Quadratic Form* is

$$ax^2 + bx + c = 0 \quad (a \neq 0).$$

◆ Any quadratic equation can have a maximum of 2 solutions.
◆ A quadratic equation can be solved graphically or algebraically.

Solving Quadratic Equations Graphically

Example

The graph of $y = x^2 - 6x + 5$ is shown.

The solutions of the quadratic equation $x^2 - 6x + 5 = 0$ are $x = 1$ and $x = 5$ because these are the x values of the points where $y = 0$ on the graph.

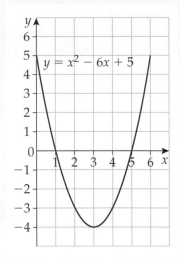

The solutions are also called the roots of the equation. The roots are the x-values where the graph crosses the x-axis.

Solving Quadratic Equations Algebraically

To solve a quadratic equation algebraically:

◆ arrange the equation in *Standard Quadratic Form*
◆ factorise the left side of the equation by deciding whether you are dealing with a common factor, a difference of two squares, or a trinomial.

The following Example considers all of the possibilities.

Example

Solve for x: a) $4x^2 - 12x = 0$ b) $x^2 - 25 = 0$
                c) $x^2 + 9x = -20$ d) $3x^2 + 16x + 5 = 0$

(Solution)

a) $4x^2 - 12x \qquad = 0$ (*no numerical term*)
     $\Rightarrow 4x\,(x - 3) \qquad = 0$ (*common factor is $4x$*)
     $\Rightarrow 4x = 0$ or $x - 3 = 0$ (*two possible solutions*)
     $\Rightarrow x = 0$ or $x = 3$.

b) $x^2 - 25 \qquad\qquad = 0$ (*difference of two squares*)
     $\Rightarrow x^2 - 5^2 \qquad\quad\, = 0$ (*write 25 as 5^2*)
     $\Rightarrow (x + 5)(x - 5) \quad = 0$ (*factorised*)
     $\Rightarrow x + 5 = 0$ or $x - 5 = 0$ (*two possible solutions*)
     $\Rightarrow x = -5$ or $x = 5$.

Example continued >

Example *continued*

c) $x^2 + 9x \qquad\qquad = -20$ *(trinomial)*
  $\Rightarrow x^2 + 9x + 20 \qquad = 0$ *(first make right side = 0)*
  $\Rightarrow (x + 4)(x + 5) \qquad = 0$ *(trinomial factorised)*
  $\Rightarrow x + 4 = 0$ or $x + 5 = 0$ *(two possible solutions)*
  $\Rightarrow x = -4$ or $x = -5.$

d) $3x^2 + 16x + 5 \qquad\quad = 0$ *(trinomial)*
  $\Rightarrow (3x + 1)(x + 5) \qquad = 0$ *(trinomial factorised)*
  $\Rightarrow 3x + 1 = 0$ or $x + 5 = 0$ *(two possible solutions)*
  $\Rightarrow 3x = -1$ or $x = -5$
  $\Rightarrow x = -\dfrac{1}{3}, x = -5.$

The Quadratic Formula

If you still cannot find factors to enable you to solve a quadratic equation, you should use the **quadratic formula**.

This formula is printed on Page 2 of your Credit paper. It is the first formula in the formulae list.

The roots of the equation $ax^2 + bx + c = 0$ are $x = \dfrac{-b \pm \sqrt{b^2 - 4ac}}{2a}$.

If the exam question asks you to 'give your answer correct to one or two decimal places', then you must use the formula. This rounding instruction is your big clue!

Example

Solve the equation $x^2 + 3x = 5$
Give your answer correct to 1 decimal place.

(Solution)

(The rounding instruction means we will use the quadratic formula. First, though, we must make the right side zero.)

Thus $x^2 + 3x - 5 = 0.$
Comparing this with $ax^2 + bx + c = 0$, we see that $a = 1$, $b = 3$ and $c = -5$.
Also $b^2 - 4ac = 3^2 - 4 \times 1 \times (-5) = 9 - (-20) = 9 + 20 = 29.$

Example *continued* ➤

Example continued

Now $\quad x = \dfrac{-b \pm \sqrt{b^2 - 4ac}}{2a}$

$\quad = \dfrac{-3 \pm \sqrt{29}}{2}$

$\quad = \dfrac{-3 \pm 5\cdot385\ldots}{2} \qquad$ (do not round yet)

$\quad = \dfrac{-3 + 5\cdot385\ldots}{2} \ $ or $\ \dfrac{-3 - 5\cdot385\ldots}{2}$

$\quad = \dfrac{2\cdot385\ldots}{2} \ $ or $\ \dfrac{-8\cdot385\ldots}{2} \qquad$ (show working)

$\quad = 1\cdot192\ldots \ $ or $\ -4\cdot192\ldots$

Hence $\quad x = 1\cdot2$ or $x = -4\cdot2$. $\qquad$ (to 1 decimal place)

For Practice

(The answers to the following questions are given in Appendix 2.)

14.1 Simplify: $\quad 3(a + 2) - 2(1 + a) + 4a$.

14.2 Simplify: $\quad (x + 1)(x^2 + 2x - 3)$.

14.3 Factorise: $\quad 4ab - 10bc$.

14.4 Factorise: $\quad$ **a)** $p^2 - 25q^2 \qquad$ **b)** $2x^2 - 7x - 4$.

14.5 Solve: $\quad 5x - 3 = 3x + 17$.

14.6 Solve: $\quad 2(x - 3) + 5(x + 2) = 25$.

14.7 Solve the pair of equations: $\qquad 3x - 2y = 7$
$\qquad\qquad\qquad\qquad\qquad\qquad\qquad\quad 4x + y = 2$.

14.8 Solve: $\quad$ **a)** $2x^2 - 50 = 0 \qquad$ **b)** $3x^2 - 5x = 2$.

14.9 Solve (giving your answer to two decimal places): $2x^2 + x - 5 = 0$.

Chapter 15

SIMPLIFYING ALGEBRAIC EXPRESSIONS

Almost all of the work in this chapter is aimed at those who are aiming for Credit Level. Those who are studying at General Level should move on to other chapters once they have read and understood the first topics on the evaluation and construction of simple formulae.

What You Should Know

a) for General Level: how to
   ◆ evaluate and construct simple formulae.

b) and in addition, for Credit Level: how to
   ◆ change the subject of a formula
   ◆ derive a formula from a graph
   ◆ simplify algebraic fractions
   ◆ work with indices and surds
   ◆ find the values of functions.

Formulae

A formula is an equation which shows the relationship between different quantities. One very important formula which you may have heard of is $E = mc^2$. (E is called the *subject* of the formula.)

Formulae occur not only in Mathematics exam questions, but are used widely in many other subjects such as Physics, Engineering and Economics.

Evaluating a Formula

Example

Two quantities a and b are connected by the formula $a = 4b - 6$. Find a when $b = 20$.

Solution:
$$a = 4b - 6$$
$$\Rightarrow a = 4 \times 20 - 6 \qquad \text{(replace } b \text{ with 20)}$$
$$\Rightarrow a = 80 - 6$$
$$\Rightarrow a = 74.$$

Constructing a Formula

As well as being able to evaluate a formula, you are sometimes required to construct a simple formula from a set of data.

Example

Two quantities r and s were measured in an experiment.
The results are as follows:

r	1	2	3	4	5	6	7	8	9	10
s	8	11	14	17						

a) Find a formula for s in terms of r.

b) Use your formula to find s when $r = 10$.

Solution:

a) After looking carefully at the data, we see that each s value is **5 more** than **3 times** the r value. (Check by finding $3r$, thus $3r + 5$.)
Hence $s = 3r + 5$

b) Since $s = 3r + 5$
$$s = 3 \times 10 + 5 \qquad \textit{(replacing r with 10)}$$
$$= 30 + 5$$
$$= 35.$$

C

Finding a Formula from a Graph

Credit candidates may be asked to find a formula or equation from a graph. (This is usually the formula for a straight line.)

Example

Use the information given in the graph to write an equation for y in terms of x.

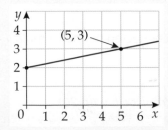

Example *continued* ➤

Example *continued*

(Solution)

Gradient of line $= \dfrac{\text{vertical height}}{\text{horizontal distance}} = \dfrac{1}{5}$.

y-intercept $= (0, 2)$.

Hence equation of line (or formula) is:

$y = \dfrac{1}{5}x + 2$.   (since $y = mx + c$)

Example

Two quantities p and q were measured in an experiment, and the results are shown in the graph. Find a formula connecting p and q.

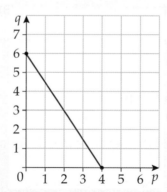

(Solution)

Gradient of line $= \dfrac{\text{vertical height}}{\text{horizontal distance}} = -\dfrac{6}{4} = -\dfrac{3}{2}$.

y-intercept $= (0, 6)$.

In terms of 'x' and 'y', the formula (or equation) is $y = -\dfrac{3}{2}x + 6$

Hence in terms of p and q, the formula (or equation) is $q = -\dfrac{3}{2}p + 6$.

Changing the Subject of a Formula

Sometimes when we are using a formula, we are given a value for the subject, and asked to find a value for one of the other terms in the formula. It is useful to be able to make *any* term the subject of the formula.

HOW TO PASS STANDARD GRADE MATHEMATICS

Example

Make x the subject of the formula $y = 2x + 5$.

(Solution)
(What we are really being asked here is to solve this equation for x.)
$$y = 2x + 5 \Rightarrow 2x + 5 = y$$
$$\Rightarrow \quad 2x = y - 5$$
$$\Rightarrow \quad x = \frac{y - 5}{2}.$$

The subject of the formula is now x.

Functions

A *function* is a mathematical relationship between two quantities, and describes how one of them depends on the other.

A function is a relationship which connects every member of one set of numbers (A) with exactly one member of another set of numbers (B).

Key Points

If 'p is a function of q', then p **depends** on q (but q does not depend on p), and we write this in functional notation as $p = f(q)$.

In this relationship, p is called the *dependent* variable, and q is the *independent* variable.

(If we are graphing a functional relationship, we almost always plot the dependent variable on the 'y'-axis, and the independent variable on the 'x'-axis.)

A function may be a simple linear relationship such as $f(x) = 4x - 1$, or a more complicated quadratic relationship such as $f(x) = x^2 + 2x - 3$.

Evaluating a Function

Example

A function $g(x)$ is defined as $g(x) = 6x + 2$.

a) Find the value of $g(10)$.

b) If $g(a) = -28$, find the value of a.

Example continued ➤

Example *continued*

(Solution)

a) $g(x) = 6x + 2$
$\Rightarrow g(10) = 6 \times 10 + 2$
$= 60 + 2$
$= 62.$

b) $g(a) = 6a + 2 = -28$
$\Rightarrow \quad 6a \quad = -28 - 2$
$= -30$
$\Rightarrow \quad a \quad = -5.$

Graphing a Function

The graph of a function can be drawn after evaluating the function for several values of the independent variable (usually x). The graph of the function $f(x) = ax + b$ is the straight line with equation $y = ax + b$.

Example

Draw graphs of the functions   a) $f(x) = 4x - 1$ b) $f(x) = x^2 + 2x - 3$.

(Solution)

a) $f(x) = 4x - 1 \Rightarrow f(0) = 4 \times 0 - 1 = -1,$ [plot $(0, -1)$]
$f(1) = 4 \times 1 - 1 = 3,$ [plot $(1, 3)$]
$f(2) = 4 \times 2 - 1 = 7.$ [plot $(2, 7)$]

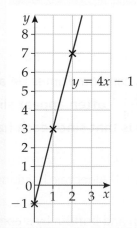

Example *continued* ➤

Example *continued*

b) $f(x) = x^2 + 2x - 3 \Rightarrow f(-4) = (-4)^2 + 2(-4) - 3 = 5$ [plot $(-4, 5)$]
$f(-3) = (-3)^2 + 2(-3) - 3 = 0$ [plot $(-3, 0)$]
$f(-2) = (-2)^2 + 2(-2) - 3 = -3$ [plot $(-2, -3)$]
$f(-1) = (-1)^2 + 2(-1) - 3 = -4$ [plot $(-1, -4)$]
$f(0) = (0)^2 + 2(0) - 3 \quad = -3$ [plot $(0, -3)$]
$f(1) = (1)^2 + 2(1) - 3 \quad = 0$ [plot $(1, 0)$]
$f(2) = (2)^2 + 2(2) - 3 \quad = 5$ [plot $(2, 5)$]

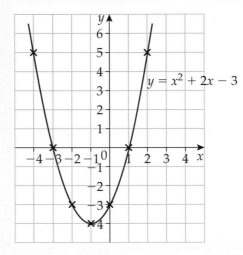

$y = x^2 + 2x - 3$

Algebraic Fractions

The appearance of algebraic fractions can sometimes be a bit alarming to candidates. Expressions such as $\dfrac{2}{y}$ or $\dfrac{x+3}{4}$ often appear slightly threatening, but the important point to remember is that since each has a numerator and a denominator, *each is a fraction*.

We revised arithmetic fractions in Chapter 3. Algebraic fractions can be treated in the same way as arithmetic fractions.

◆ They may be simplified by dividing top and bottom by a common factor.

◆ They may be added, subtracted, multiplied and divided using the same rules as before.

Simplifying Algebraic Fractions

Remember

In arithmetic $\dfrac{10}{12} = \dfrac{2 \times 5}{2 \times 6} = \dfrac{2}{2} \times \dfrac{5}{6} = 1 \times \dfrac{5}{6} = \dfrac{5}{6}$.

In effect we are dividing top and bottom by 2, and we might show this more simply as

$$\frac{\cancel{10}}{\cancel{12}} = \frac{5}{6}.$$

The following illustrations show how algebraic fractions may be simplified by factorisation and cancellation.

Example

a) $\dfrac{2}{6x}$ $= \dfrac{\cancel{2} \times 1}{\cancel{2} \times 3 \times x}$ $= \dfrac{1}{3x}$

b) $\dfrac{xy}{5y}$ $= \dfrac{x \times \cancel{y}}{5 \times \cancel{y}}$ $= \dfrac{x}{5}$

c) $\dfrac{8a^2}{6a}$ $= \dfrac{\cancel{2} \times 4 \times \cancel{a} \times a}{\cancel{2} \times 3 \times \cancel{a}}$ $= \dfrac{4a}{3}$

d) $\dfrac{(x+4)^2}{3(x+4)}$ $= \dfrac{(x+4)\cancel{(x+4)}}{3 \times \cancel{(x+4)}}$ $= \dfrac{(x+4)}{3}$

e) $\dfrac{x^2 - 9}{2x^2 + 5x - 3}$ $= \dfrac{\cancel{(x+3)}(x-3)}{(2x-1)\cancel{(x+3)}}$ $= \dfrac{(x-3)}{(2x-1)}$.

Adding, Subtracting, Multiplying and Dividing

Key Points

◆ As was the case in arithmetic, fractions may only be added or subtracted **when they have a common denominator**.

◆ Often, the safest way to find a common denominator is to multiply the two denominators.

◆ After adding or subtracting, the final fraction should be simplified.

◆ When multiplying fractions, the numerators are multiplied, then the denominators are multiplied, and the final fraction is then simplified.

◆ When dividing fractions, the first fraction is **multiplied by the inverse of the**

The following illustrations demonstrate the four operations with algebraic fractions.

Example

a) (addition)

Since $\dfrac{2}{7} + \dfrac{3}{7} = \dfrac{5}{7}$ then $\dfrac{2}{x} + \dfrac{3}{x} = \dfrac{5}{x}$.

b) (subtraction)

Since $\dfrac{2}{5} - \dfrac{1}{3} = \dfrac{6}{15} - \dfrac{5}{15} = \dfrac{1}{15}$,

then $\dfrac{2}{a} - \dfrac{1}{b} = \dfrac{2b}{ab} - \dfrac{a}{ab} = \dfrac{2b - a}{ab}$.

c) (multiplication)

Since $\dfrac{5}{9} \times \dfrac{2}{7} = \dfrac{10}{63}$ then $\dfrac{r}{s} \times \dfrac{t}{v} = \dfrac{rt}{sv}$.

d) (division)

Since $\dfrac{2}{3} \div \dfrac{4}{5} = \dfrac{2}{3} \times \dfrac{5}{4} = \dfrac{10}{12} = \dfrac{5}{6}$

then $\dfrac{2s}{3t} \div \dfrac{4s^2}{5t} = \dfrac{2s}{3t} \times \dfrac{5t}{4s^2} = \dfrac{10st}{12s^2 t} = \dfrac{5}{6s}$.

C Indices

In Chapter 4, we made use of indices in writing scientific notation like $3 \cdot 2 \times 10^4$. Indices are used frequently as a simple way of expressing numbers, and you will already know that, for example, 3^4 means $3 \times 3 \times 3 \times 3$. We read 3^4 as '3 to the power 4'. The number '3' is called the **base**, and the number '4' is called the **index** (or the power).

Here we look at how to simplify terms involving indices. If in a problem we meet two index terms which we wish to combine, we can only do so provided they have the same base.

We look at indices which are positive integers, negative integers, and fractions. We also look at the meaning of a zero index!

Remember

The rules of indices are not printed in the formulae list. They must be learned!

1   $a^m \times a^n = a^{m+n}$

2   $a^m \div a^n = a^{m-n}$

Remember continued ➤

Remember *continued*

3 $(a^m)^n = a^{m \times n}$

4 $a^0 = 1$

5 $a^{-m} = \dfrac{1}{a^m}$

6 $a^{\frac{m}{n}} = \sqrt[n]{a^m}$

The rules are illustrated in the following examples.

Example

a) $6^2 \times 6^7 = 6^9$ also $a^4 \times a^3 = a^7$ (Rule 1)

b) $5^8 \div 5^2 = 5^6$ also $b^{10} \div b^8 = b^2$ (Rule 2)

c) $(3^6)^2 = 3^{12}$ also $(x^2)^4 = x^8$ (Rule 3)

d) $8^0 = 1$ also $y^0 = 1$ (Rule 4)

e) $4^{-5} = \dfrac{1}{4^5}$ also $c^{-6} = \dfrac{1}{c^6}$ (Rule 5)

f) $8^{\frac{2}{3}} = \sqrt[3]{8^2}$

       $= \sqrt[3]{64}$

       $= 4$ also $a^{\frac{3}{4}} = \sqrt[4]{a^3}$ (Rule 6)

Key Points

◆ In the illustration of Rule 6, $\sqrt[3]{64}$ means 'the cube root of 64'. (This is 4, since $4 \times 4 \times 4 = 64$.)

◆ A cube root may also be written as, for example, $8^{\frac{1}{3}}$.

(This is because $8^{\frac{1}{3}} \times 8^{\frac{1}{3}} \times 8^{\frac{1}{3}} = 8^{\frac{1}{3} + \frac{1}{3} + \frac{1}{3}} = 8^1 = 8$ – using Rule 1)

Hence $8^{\frac{1}{3}} = 2$.

◆ Similarly, a square root might be written as, for example, $25^{\frac{1}{2}}$, and a fourth root as $81^{\frac{1}{4}}$.

Example

Simplify **a)** $t^{\frac{1}{2}}(t^{-\frac{1}{2}} + t^{\frac{1}{2}})$ **b** $\dfrac{y^{\frac{2}{3}} \times y^{\frac{1}{4}}}{y^{\frac{5}{6}}}$.

(Solution)

a) $t^{\frac{1}{2}}(t^{-\frac{1}{2}} + t^{\frac{1}{2}})$

$= t^0 + t^1$ (*expanding bracket and using Rule 1*)

$= 1 + t$. (*using Rule 4*)

b) $\dfrac{y^{\frac{2}{3}} \times y^{\frac{1}{4}}}{y^{\frac{5}{6}}} = \dfrac{y^{\frac{2}{3} + \frac{1}{4}}}{y^{\frac{5}{6}}}$ (*Rule 1 – add indices*)

$= \dfrac{y^{\frac{8}{12} + \frac{3}{12}}}{y^{\frac{5}{6}}} = \dfrac{y^{\frac{11}{12}}}{y^{\frac{5}{6}}} = \dfrac{y^{\frac{11}{12}}}{y^{\frac{10}{12}}}$

$= y^{\frac{11}{12} - \frac{10}{12}} = y^{\frac{1}{12}}$. (*Rule 2 – subtract indices*)

Hints and Tips

If you were asked to evaluate $4^3 \times 3^2$, it would be incorrect to write this as $(4 \times 3)^{3+2}$ i.e. as 12^5 ($= 248\,832$).

You cannot apply the rules of indices (Rule 1 here) *to terms with different bases*. (In this case, you would have to calculate $4^3 \times 3^2$ as 64×9 ($= 576$).)

C ## Surds

Key Words and Definitions

There are many *number systems* used in Mathematics, some of which you are quite familiar with:

◆ N – the set of natural numbers $\{1, 2, 3, \ldots\}$
◆ W – the set of whole numbers $\{0, 1, 2, 3, \ldots\}$
◆ Z – the set of integers $\{\ldots, -2, -1, 0, 1, 2, 3, \ldots\}$

Key Words and *Definitions* continued ➢

Key Words and Definitions continued

There is also a set of *rational numbers*, Q, which can be written as *the ratio of two integers*. Examples of rational numbers are:

a) $5\left(=\dfrac{5}{1}\right)$ b) $0{\cdot}3\left(=\dfrac{3}{10}\right)$ c) $0{\cdot}66666...\left(=\dfrac{2}{3}\right)$.

Numbers which cannot be written as the ratio of two integers are called *irrational numbers*.

A very important example of an irrational number is π.

Some of the irrational numbers are called SURDS. These are the *irrational roots* of rational numbers. Examples of surds are $\sqrt{2}(=1{\cdot}414...)$, $\sqrt{3}(=1{\cdot}732...)$. Note, however, that $\sqrt{4}$ is not a surd, because $\sqrt{4}=2$, which is rational.

Simplifying Surds

Key Points

1 Suppose we wish to evaluate $\sqrt{n}$, and that n can be written as a product of factors $a \times b$;

$$\text{then } \sqrt{n} = \sqrt{a \times b} = \sqrt{a} \times \sqrt{b}.$$

(When we look for a pair of factors, we look for one which is the highest perfect square).

2 $\sqrt{\dfrac{a}{b}} = \dfrac{\sqrt{a}}{\sqrt{b}}$.

Example

Simplify a) $\sqrt{28}$ b) $\sqrt{200}$ c) $\sqrt{\dfrac{9}{16}}$ d) $\sqrt{\dfrac{50}{72}}$.

(Solution)

a) $\sqrt{28} = \sqrt{4 \times 7} = \sqrt{4} \times \sqrt{7} = 2\sqrt{7}$.

b) $\sqrt{200} = \sqrt{100 \times 2} = \sqrt{100} \times \sqrt{2} = 10\sqrt{2}$.

c) $\sqrt{\dfrac{9}{16}} = \dfrac{\sqrt{9}}{\sqrt{16}} = \dfrac{3}{4}$.

d) $\sqrt{\dfrac{50}{72}} = \dfrac{\sqrt{50}}{\sqrt{72}} = \dfrac{\sqrt{25 \times 2}}{\sqrt{36 \times 2}} = \dfrac{\sqrt{25} \times \sqrt{2}}{\sqrt{36} \times \sqrt{2}} = \dfrac{5\sqrt{2}}{6\sqrt{2}} = \dfrac{5}{6}$.

Adding and Subtracting Surds

If surds are in the form of 'like terms', then they can be gathered by addition and subtraction just like any other 'like terms' in Algebra.

Example

Simplify a) $\sqrt{7} + \sqrt{7}$ b) $8\sqrt{11} - 5\sqrt{11}$ c) $\sqrt{18} + \sqrt{50}$

(Solution)

a) $\sqrt{7} + \sqrt{7} = 2\sqrt{7}$.

b) $8\sqrt{11} - 5\sqrt{11} = 3\sqrt{11}$

c) $\sqrt{18} + \sqrt{50} = \sqrt{9 \times 2} + \sqrt{25 \times 2} = 3\sqrt{2} + 5\sqrt{2} = 8\sqrt{2}$.

Rationalising a Denominator

A term with a surd in the denominator is difficult to deal with. We can remove the surd however by *rationalising the denominator*.

Example

Simplify: $\dfrac{4}{\sqrt{10}}$.

(Solution)

$$\frac{4}{\sqrt{10}} = \frac{4}{\sqrt{10}} \times \frac{\sqrt{10}}{\sqrt{10}} = \frac{4\sqrt{10}}{10} = \frac{2\sqrt{10}}{5}.$$

(Multiplying top and bottom by $\sqrt{10}$ does not change the original term, because $\dfrac{\sqrt{10}}{\sqrt{10}} = 1$.

Also, in the bottom line $\sqrt{10} \times \sqrt{10} = \sqrt{100} = 10$.)

Example

Rationalise the denominator for the expression $\dfrac{1}{(\sqrt{3} + 1)}$.

(Solution)

(Here we multiply top and bottom by what is called the *conjugate* of the denominator. It is written as $(\sqrt{3} - 1)$.)

Hence $\dfrac{1}{(\sqrt{3} + 1)} = \dfrac{1}{(\sqrt{3} + 1)} \times \dfrac{(\sqrt{3} - 1)}{(\sqrt{3} - 1)} = \dfrac{(\sqrt{3} - 1)}{3 + \sqrt{3} - \sqrt{3} - 1)} = \dfrac{\sqrt{3} - 1}{2}$.

For Practice

(The answers to the following questions are given in Appendix 2.)

15.1 The quantities a, t, u, and v are connected by the formula $v = u + at$.
Find the value of v, when $u = 2$, $a = 1\cdot5$, and $t = 7$.

15.2 Change the subject of the formula $v^2 = u^2 + 2as$ to a.

15.3 A function $h(x)$ is defined by $h(x) = 2x^2 - 1$.

    a) Find the value of $h(3)$.

    b) If $h(t) = 97$, find the values of t.

15.4 Simplify: $\dfrac{2x^2 - x - 3}{2x^2 + x - 6}$.

15.5 Find   **a)** $\dfrac{2}{x} + \dfrac{x}{2}$ **b)** $\dfrac{4p}{3q^2} \div \dfrac{2p^2}{9q}$.

15.6 Simplify: $a^{-\frac{1}{2}}\left(a^{\frac{3}{2}} - a^{\frac{1}{2}}\right)$.

15.7 Simplify: **a)** $\sqrt{63}$ **b)** $\dfrac{10}{\sqrt{12}}$ **c)** $\sqrt{50} + \sqrt{2} - \sqrt{18}$.

TRIGONOMETRY I: CALCULATIONS WITH TRIANGLES

Trigonometry is quite simply the study of triangles. It is concerned with the relationships between the lengths of the sides of the triangle, and the sizes of the angles in the triangle. It makes use of functions which are special properties of the angles. These are called sine, cosine and tangent.

Key Words

Opposite ★ Adjacent ★ Hypotenuse ★ Sine ★ Cosine ★ Tangent ★ Sine Rule ★ Cosine Rule

What You Should Know

a) for General Level: how to
   ◆ calculate the length of a side or the size of an angle in a right-angled triangle.

b) and in addition, for Credit Level: how to
   ◆ use the sine rule and cosine rule in a scalene (non-right-angled) triangle
   ◆ calculate the area of a scalene triangle.

Trigonometry and Right-angled Triangles

In Chapter 12, we met the Theorem of Pythagoras. This theorem tells us the connection between the sizes of the three sides of a right-angled triangle, so that, knowing two of them, we can calculate the third.

If we know the size of a side and the size of an angle in a right-angled triangle then, by using trigonometry, we can calculate the sizes of the other sides, by making use of the trigonometric ratios. Or if we know the sizes of two sides, we can calculate the sizes of the angles (and of course, the third side).

Key Points

◆ the *hypotenuse* is opposite the right angle
◆ the *opposite* is the side facing the angle that is marked
◆ the *adjacent* is the remaining side.

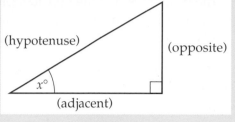

$$\sin x° = \frac{\text{opposite}}{\text{hypotenuse}}; \quad \cos x° = \frac{\text{adjacent}}{\text{hypotenuse}}; \quad \tan x° = \frac{\text{opposite}}{\text{adjacent}}$$

('S O H')   ('C A H')   ('T O A')

These relationships are printed in the formulae list of the General paper. They are not printed in the formulae list of the Credit paper.

Calculating the Length of a Side in a Right-angled Triangle

Example

The diagram shows a rectangular metal framework. The diagonal strut AC is 8 cm long, and angle ACB = 42°.
Calculate the length of AB.
(Give your answer correct to two decimal places.)

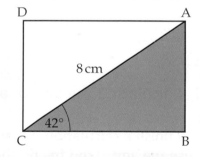

(Solution)
(First draw triangle ABC and label its sides.)

Searching: S O H   C A H   T O A
              ? ✓   ✓   ?

Hence use: $\sin 42° = \dfrac{\text{opp}}{\text{hyp}} \Rightarrow 0{\cdot}669 = \dfrac{\text{AB}}{8}$

Hence AB $= 0{\cdot}669 \times 8 = 5{\cdot}353$

Hence AB $= 5{\cdot}35$ cm (to two dec. places).

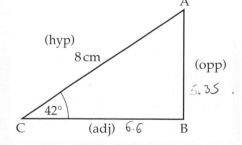

◆ **Always check that your calculator is in degree mode.**
◆ **Wait till the finish of the calculation before you round any values.**

C

Example

The diagram shows the sloping roof section of part of a building. The roof makes an angle of 65° with the wall of the building, and the length of RT is 7·4 metres.
Calculate the length of the roof RS.

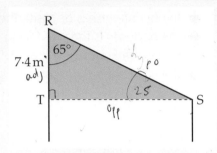

(Solution)
(Again, 'isolate' the triangle and label its sides.)

Searching: S O H   C A H   T O A   (adj) 7·4 m
              ?   ✓ ?   ✓

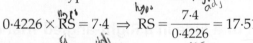

Hence use: $\cos 65° = \dfrac{\text{adj}}{\text{hyp}} \Rightarrow 0·4226 = \dfrac{7·4}{\text{RS}}$

Hence $0·4226 \times \text{RS} = 7·4 \Rightarrow \text{RS} = \dfrac{7·4}{0·4226} = 17·51$

The length of the sloping roof is thus 17·51 metres.

Calculating the Size of an Angle in a Right-angled Triangle

!

To finish a calculation to find the size of an angle, you will need to be able to use the 'Inv', '2nd fn', or 'shift' button of your calculator (depending on which calculator you have).

For Practice

Use your calculator to check the following results:

       $\sin x° = 0·927$ $\cos y° = 0·542$ $\tan z° = 1·064$

      $\Rightarrow x = 68·0$ $\Rightarrow y = 57·2$ $\Rightarrow z = 46·8.$

Example

The diagram represents a piece of glass in the shape of a right-angled triangle.
Find the size of the angle marked $x°$.

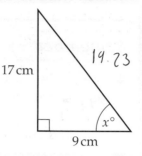

17 cm

19.23

9 cm

$x°$

(Solution)

(The second diagram shows the triangle fully labelled.)

Searching: S O H   C A H   T O A
              ✓   ✓   ✓ ✓

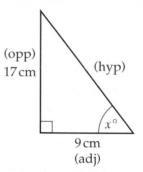

(opp)
17 cm

(hyp)

9 cm
(adj)

$x°$

Hence use: $\tan x° = \dfrac{\text{opp}}{\text{adj}} \;\Rightarrow\; \tan x° = \dfrac{17}{9}.$

Hence $\tan x° = 1\cdot888\ldots$ *(avoid rounding at this stage)*

             $\Rightarrow\; x = 62\cdot1$ *(using Inv tan key)*

C Trigonometry and Non Right-angled Triangles

When a triangle does not have a right angle, the three ratios can not be used. The methods used instead are the Sine Rule, and the Cosine Rule.

Key Points

The Sine Rule: In triangle ABC:

$$\frac{a}{\sin A} = \frac{b}{\sin B} = \frac{c}{\sin C}.$$

The Cosine Rule: In triangle ABC:

$$a^2 = b^2 + c^2 - 2bc\cos A \quad \text{or}$$

$$\cos A = \frac{b^2 + c^2 - a^2}{2bc}.$$

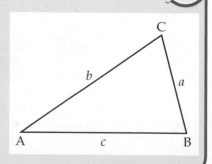

Which Rule to Use?

> **Remember**
>
> ◆ Use the Sine Rule when you know the sizes of **one side and the angle opposite it** (along with, say, the size of another side or angle).
> ◆ Use the Cosine Rule (first form) when you know the sizes of **two sides and their included angle**.
> ◆ Use the Cosine Rule (second form) when you know the sizes of **all three sides**.

Using the Sine Rule

Calculating the length of a side

Example

Find the length of BC in triangle ABC.
Give your answer correct to two decimal places.

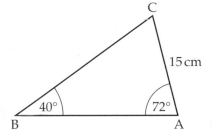

(Solution)

(Here we know the length of one side (15 cm), the size of the angle opposite it (40°), and another angle.)

The second diagram shows the triangle fully labelled.

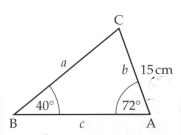

Searching: $\dfrac{a}{\sin A} = \dfrac{b}{\sin B} = \dfrac{c}{\sin C}$.

Hence: $\dfrac{a}{\sin 72°} = \dfrac{15}{\sin 40°}$.

Hence $a = \dfrac{15 \times \sin 72°}{\sin 40°} = 22{\cdot}1937\ldots$

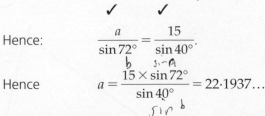

Hence BC = 22·19 cm.

Calculating the size of an angle

Example

Calculate the size of angle ABC in triangle ABC.

(Solution)

(Here we know the length of one side (28 cm), the size of the angle opposite it (64°), and another side.)

The second diagram shows the triangle fully labelled.

Searching:  $\dfrac{a}{\sin A} = \dfrac{b}{\sin B} = \dfrac{c}{\sin C}$.

Hence: $\dfrac{28}{\sin 64} = \dfrac{30}{\sin B}$.

Hence: $\sin B = \dfrac{30 \times \sin 64}{28} = 0.96299\ldots$ (*do not round yet*)

$\Rightarrow B = 74.36°.$ (*using Inv sin key*)

Hence angle ABC $= 74.36°.$

Using the Cosine Rule

Calculating the length of a side

Example

The diagram shows the design of a small flag ABC.
A piece of gold ribbon is to be attached to side BC.
What length of ribbon is required?
(Give your answer to three significant figures.)

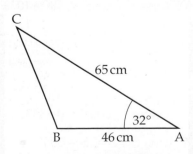

Example *continued*

Example *continued*

(Solution)

(Here we know sizes of two sides and their included angle, and so use the first form of the cosine rule.)

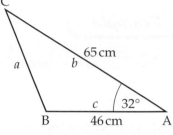

The second diagram shows the triangle fully labelled.

$a^2 = b^2 + c^2 - 2bc \cos A \Rightarrow a^2 = 65^2 + 46^2 - 2 \times 65 \times 46 \times \cos 32°$

$= 4225 + 2116 - 5071 \cdot 328$

$= 1269 \cdot 672.$ (*do not round yet*)

Hence $a = \sqrt{1269 \cdot 672} = 35 \cdot 632 = 35 \cdot 6.$ (*to 3 sig. figs.*)

The length of ribbon needed is thus $35 \cdot 6$ cm.

Calculating the size of an angle

Example

The diagram shows triangle RST.
Calculate the size of angle RST.

(Solution)

(Here we know sizes of all three sides, and so we use the second form of the cosine rule.)

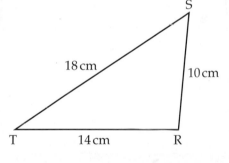

The second diagram shows the triangle fully labelled.

Here $\quad \cos S = \dfrac{r^2 + t^2 - s^2}{2rt}$

$= \dfrac{18^2 + 10^2 - 14^2}{2 \times 18 \times 10}$

$= \dfrac{324 + 100 - 196}{360}$

$= \dfrac{228}{360}.$

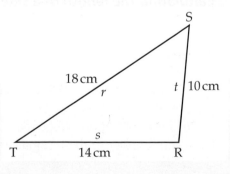

Hence $\quad \cos S = 0 \cdot 6333\ldots$ (*do not round yet*)

$\Rightarrow S = 50 \cdot 7°$ (*using Inv cos key*)

Hence angle RST $= 50 \cdot 7°.$

Sometimes you may be required to apply the Sine Rule or the Cosine Rule to a triangle which occurs in a question involving *three-figure bearings*. (Three-figure bearings were introduced towards the end of Chapter 9.)

In the Credit paper, a trigonometry question may require you to use right-angled triangle trigonometry and Sine rule or Cosine rule!

Example

An observer at B measures the angle of elevation of a kite at A to be 72°.

A second observer at C measures the angle of elevation of the kite at A to be 43°.

The distance between the observers is 50 m.

Find the vertical height of the kite above the ground. (Round your answer to 3 sig. figs.)

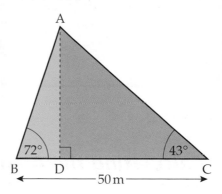

(Solution)
(The second diagram shows triangle ABC fully labelled.
Angle BAC is found to be 65° (since BAC = 180° − (72 + 43)°.)

Searching:
$$\overset{\checkmark}{\frac{a}{\sin A}} = \overset{?}{\frac{b}{\sin B}} = \overset{?}{\frac{c}{\sin C}}.$$
$$\underset{\checkmark}{} \quad \underset{\checkmark}{} \quad \underset{\checkmark}{}$$

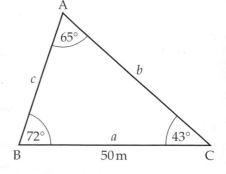

Choosing:
$$\frac{a}{\sin A} = \frac{c}{\sin C},$$

then
$$\frac{50}{\sin 65°} = \frac{c}{\sin 43°}.$$

Hence
$$c = \frac{50 \times \sin 43°}{\sin 65°} = 37 \cdot 6.$$

Hence $AB = 37 \cdot 6$ m.

Now we use the right-angled triangle ABD, fully labelled as shown.

Searching: $\overset{?}{S}\overset{\checkmark}{O}H \quad \overset{\checkmark}{C}A\overset{?}{H} \quad TOA$

Hence use: $\sin 72° = \dfrac{\text{opp}}{\text{hyp}} \Rightarrow 0 \cdot 9510\ldots = \dfrac{AD}{37 \cdot 6}$

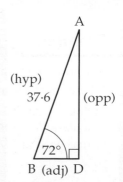

Hence $AD = 37 \cdot 6 \times 0 \cdot 9510 = 35 \cdot 8$ (*to 3 sig. figs.*)

The kite is therefore $35 \cdot 8$ m vertically above the ground.

Remember

The **angle of elevation** is the angle measured upward from the horizontal.
(Point P has an angle of elevation of 26°.)

The **angle of depression** is measured downward from the horizontal.
(Point Q has an angle of depression of 18°.)

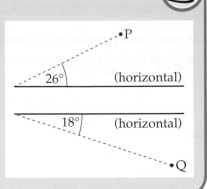

Calculating the Area of a Non Right-angled Triangle

C

Key Points

In Chapter 10, we saw that the area of a non right-angled triangle may be calculated provided we know the sizes of the base and the 'height'.

If these sizes are **not** known but you do know the sizes of **two sides and their included angle**, then the area of the triangle may be calculated from the formula

$$\text{Area} = \frac{1}{2}ab\sin C.$$

(This formula is provided in the Credit paper.)

Example

A triangular metal plate has two sides of lengths 4·2 cm, and 6·9 cm. The angle between these sides is 36°.
Calculate the area of the piece of metal.
Round your answer to three significant figures.

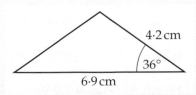

Example continued ➤

Example *continued*

(Solution)
(The second diagram shows the triangle fully labelled.)

Area $= \frac{1}{2}ab\sin C$

$= \frac{1}{2} \times 4\cdot2 \times 6\cdot9 \times \sin36°$

$= 8\cdot517\ldots$

$= 8\cdot52$ (to 3 sig. figs.).

The area is thus $8\cdot52\,\text{cm}^2$.

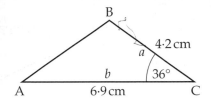

For Practice

(The answers to the following questions are given in Appendix 2.)

16.1 The diagram shows the design of a concrete ramp.

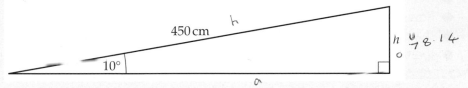

The length of the slope is 450 cm, and the slope makes an angle of 10° with the horizontal. Find the vertical height, h, of the ramp.

16.2 Triangle ABC is as shown in the diagram:

Using the information given in the diagram, find the length of BC and the sizes of the angles BAC and BCA.

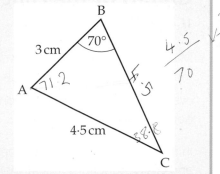

16.3 The diagram shows a plot of land LMN.
Using the information given in the diagram, find:
a) the perimeter of the plot
b) the area of the plot.

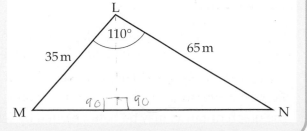

TRIGONOMETRY II: GRAPHS, EQUATIONS AND IDENTITIES

i We now extend and apply our understanding of trigonometry by looking first at the graphs of the trigonometric functions. These then allow us to tackle *trigonometric equations*. (Equations involving trigonometric terms occur frequently in Physics and Engineering.)

! This chapter is concerned purely with Credit Level topics.

What You Should Know

for Credit Level:
◆ the meanings of symmetry and periodicity
◆ the nature of the graphs of sin x, cos x, and tan x
◆ how to plot graphs *related* to the graphs of the basic functions
◆ how to solve simple equations
◆ the two important identities.

C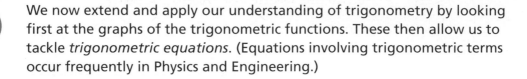

Graphs of the Functions $y = sin\ x°$, $y = cos\ x°$, and $y = tan\ x°$

Many things in our lives are *periodic*. This means that they have a pattern which repeats itself at regular intervals. Examples include the phases of the moon (repeated **i** every $29\frac{1}{2}$ days), and the motion of the minute hand of a clock (repeated every hour).

In mathematics, the sine, cosine, and tangent functions illustrate *periodicity*. The sine and cosine functions repeat themselves every 360°, and the tangent function repeats itself every 180°.

The graphs of the functions may be plotted using your calculator to find the values of the functions corresponding to selected values for x.

The graphs of sin $x°$ and cos $x°$ may be plotted fairly quickly by selecting x values of (0, 30, 60, 90, 120, …360)°.

The graph of tan $x°$ may be better plotted by selecting x values of (0, 15, 30, 45, 60, 75, 85)° and (95, 105, 120, 135, 150, 165, 180)°.

The graphs then appear as follows

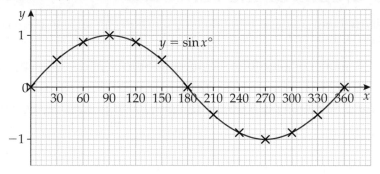

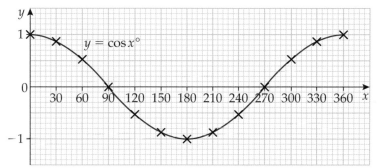

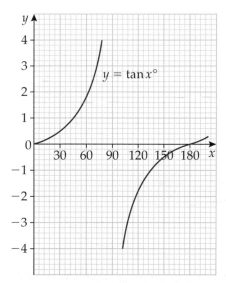

Key Points

◆ Note that the maximum and minimum values of both sine and cosine functions are ± 1.

◆ The period of both sine and cosine functions is $360°$.

◆ The period of the tangent function is $180°$.

◆ Note the symmetry in the graphs. For example $\sin 30° = \sin(180 - 30)° = \sin 150°$.

C Related Graphs of the Functions

Sometimes in problems, a basic trigonometric function can appear in a more complex form.

By graphing the more complex form, however, the problem can often be more easily understood and solved.

Here we look at a few examples of more complex forms of the sine function. These are:

◆ $y = 3 \sin x°$

◆ $y = \sin 2x°$

◆ $y = 1 + 2 \sin x°$.

$y = 3 \sin x°$

The graph may be drawn after completing a table of values:

$x°$	0	30	60	90	120	etc.					
$\sin x°$	0	0·5	0·87	1	0·87	etc.					
$3 \sin x°$	0	1·5	2·61	3	2·61	etc.					

The graph then appears as follows:

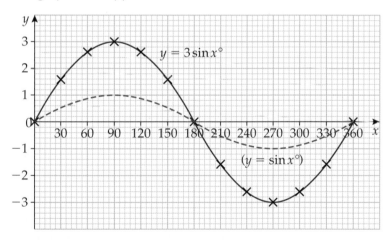

$y = \sin 2x°$

First we set up a table of values:

$x°$	0	15	30	45	60	etc.					
$2x°$	0	30	60	90	120	etc.					
$\sin 2x°$	0	0·5	0·87	1	0·87	etc.					

The graph then appears as follows:

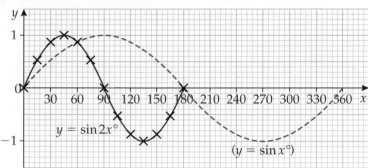

$y = 1 + 2 \sin x°$

First we set up a table of values:

$x°$	0	30	60	90	120	etc.				
$\sin x°$	0	0·5	0·87	1	0·87	etc.				
$2 \sin x°$	0	1	1·74	2	1·74	etc.				
$1 + 2 \sin x°$	1	2	2·74	3	2·74	etc.				

The graph then appears as follows:

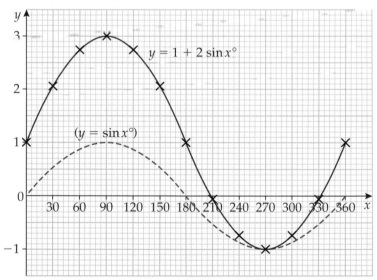

Key Points continued ➤

Key Points

◆ When a simple sine (or cosine) function is multiplied by an 'amplitude factor', so that $y = \sin x°$ becomes $y = a \sin x°$, then the maximum and minimum values become $\pm a$.

Key Points *continued*

◆ When a simple sine (or cosine) function is given a 'period factor', so that $y = \sin x°$ becomes $y = \sin bx°$, then the period of the function becomes $\frac{360°}{b}$.

◆ When a sine (or cosine) function has a constant numerical term added to it, so that $y = \sin x°$ becomes $y = c + \sin x°$, then the original graph is displaced by c vertically.

For Practice

Using your calculator to give you sets of values, plot the following functions on a sheet of graph paper: $y = 2 \cos x°$, $y = \cos 2x°$, $y = 2 + \cos x°$.

Example

Write down the equations of the following graphs.

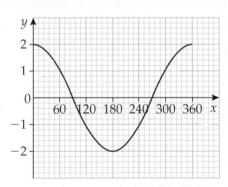

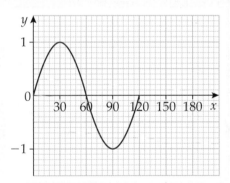

(Solution)

a) This is a cosine function.

  Its amplitude is 2.

  Its period is 360°.

  Hence $y = 2 \cos x°$.

b) This is a sine function.

  Its amplitude is 1.

  Its period is 120°.

  Hence $y = \sin 3x°$.

C Solving Trigonometric Equations

Example

Solve for x: a) $\sin x° + 1 = 1.6$ b) $5\cos x° + 2 = 1$ (both for $0 \leqslant x < 360$).

(Solution)

a) $\sin x° + 1 = 1.6 \Rightarrow \sin x° = 0.6$

   Hence $x = 36.9$.   *(using Inv sin key)*

   However, the symmetry of the upper
   part of the sine graph shows that
   $(180 - 36.9)°$ is also a solution.

   Hence $x = 36.9$ or 143.1.

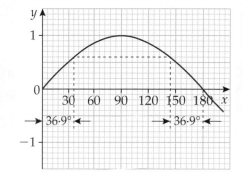

b) $5\cos x° + 2 = 1 \Rightarrow 5\cos x° = -1$

   $\Rightarrow \quad \cos x° = -\dfrac{1}{5}$

   $\Rightarrow \quad \cos x° = -0.2$.

Hence $x = 101.5$.   *(using Inv cos key)*

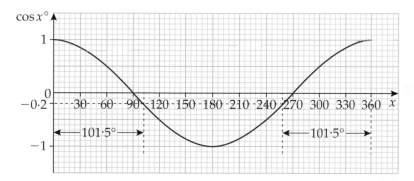

However, the symmetry of the entire cosine graph shows that $(360 - 101.5)°$ is also a solution.

Hence $x = 101.5$ or 258.5.

C *Trigonometric Identities*

The three functions, sine, cosine, and tangent, may be connected by two simple formulae.
These formulae are called Identities, and they are true for all values of angles.

> ### Remember
>
> The formulae are:
>
> $$\tan A = \frac{\sin A}{\cos A} \quad \text{and} \quad \sin^2 A + \cos^2 A = 1$$
>
> (These formulae are not printed in the Credit paper.)

i These Identities may be proved simply using a right-angled triangle:

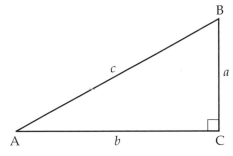

You will see that $\sin A \left(= \dfrac{\text{opp}}{\text{hyp}} \right) = \dfrac{a}{c}$, and $\cos A \left(= \dfrac{\text{adj}}{\text{hyp}} \right) = \dfrac{b}{c}$.

Hence $\dfrac{\sin A}{\cos A} = \dfrac{\frac{a}{c}}{\frac{b}{c}} = \dfrac{a}{c} \times \dfrac{c}{b} = \dfrac{a}{b}$.

Also $\tan A \left(= \dfrac{\text{opp}}{\text{adj}} \right) = \dfrac{a}{b}$.

Hence $\tan A = \dfrac{\sin A}{\cos A}$.

> ### For Practice
>
> Can you prove the second Identity ($\sin^2 A + \cos^2 A = 1$) using the same triangle?

Example

If $\sin x = \dfrac{4}{5}$, calculate the value of $\cos x$

(Solution)

$\sin^2 x + \cos^2 x = 1$

Hence $\cos^2 x = 1 - \sin^2 x$

$$= 1 - \left(\frac{4}{5}\right)^2$$

$$= 1 - \frac{16}{25}$$

$$= \frac{25}{25} - \frac{16}{25}$$

$$= \frac{9}{25}.$$

Hence $\cos x = \dfrac{3}{5}$.

Example

Simplify: $\quad \sin x \cos x \tan x$.

(Solution) $\quad \sin x \cos x \tan x = \dfrac{\sin x}{1} \times \dfrac{\cos x}{1} \times \dfrac{\sin x}{\cos x}$

$$= \sin^2 x.$$

For Practice

(The answers to the following questions are given in Appendix 2.)

C 17.1 Sketch a graph of the function $y = 2 \sin 3x°$.

C 17.2 Solve for x: $3 \sin x° = 2$ $\quad$ (for $0 \leqslant x < 360$)

C 17.3 Prove that $\cos^2 x \tan^2 x = 1 - \cos^2 x$.

STATISTICAL 'AVERAGES' AND SPREAD

We are presented with statistical information every day: on television, in newspapers, and on the Internet. It is often easy to do the arithmetic involved in statistical calculations, but sometimes more difficult to interpret the results. There is little point in being able to calculate an average to three decimal places if we do not understand how the results overall are spread.

This chapter will help you to make sense of simple everyday statistics.

What You Should Know

a) for General Level: how to
   ◆ find the mode, median, mean, and range of a data set
   ◆ deal with a frequency table
   ◆ interpret calculated statistics.

b) and in addition, for Credit Level: how to
   ◆ find the quartiles and semi-interquartile range of a data set
   ◆ calculate standard deviation.

Key Words

Mean ★ **Median** ★ **Mode** ★ **Range** ★ **Frequency Table** ★ **Quartiles** ★ **Semi-Interquartile Range** ★ **Standard Deviation** ★ **Cumulative Frequency**

The 'Averages' – Mean, Median, and Mode

There are three different ways of finding a statistical 'average'.

Remember

$$\text{Mean} = \frac{\text{sum of statistics}}{\text{number of statistics}}$$

Median = the **middle** statistic, when the statistics are **in ascending order**

Mode = the statistic which occurs **most often**.

Example

Find the mean, median, and mode of the following data sets:
a) 2   4   5   7   8   8   8
b) 51   36   43   40   55   36.

(Solution)

a) Mean $= \dfrac{2+4+5+7+8+8+8}{7} = \dfrac{42}{7} = 6.$

Median: the statistics are already in ascending order:

   2   4   5   <u>7</u>   8   8   8

Hence median (middle statistic) $= 7$.

Mode: 8   (occurs three times).

b) Mean $= \dfrac{51+36+43+40+55+36}{6} = \dfrac{261}{6} = 43{\cdot}5.$

Median: first arrange statistics in ascending order:

   36   36   40   43   51   55

(since there is no middle number, we average the numbers on either side of the centre)

Hence Median $= \dfrac{40+43}{2} = \dfrac{83}{2} = 41{\cdot}5.$

Mode: 36   (occurs twice).

ⓘ The three 'averages' do not necessarily yield the same result for a data set. In the Example,
a) mean, median, mode $= 6, 7, 8$
b) mean, median, mode $= 43{\cdot}5, 41{\cdot}5, 36$

It is therefore important to say which figure you are quoting, and say 'median =', rather than just 'average ='.

Finding the Mean, Median, and Mode from a Frequency Table

Larger sets of data are organised into frequency tables. Using such a table, it is not difficult to find the values for mean, median, and mode.

Example

The ages (to the nearest year) of children placed in a supermarket crèche one day are shown in the following frequency table.

Calculate values for the mean and the median ages. State the modal age.

age	number of children
1	3
2	4
3	9
4	10
5	12
6	2

(Solution)

We represent the age by x, the number of children (the frequency) by f, and add a third column (fx) to the table.

(This saves calculating the sum of ages as $1 + 1 + 1 + 2 + 2 + 2 + 2 + 3 + 3 + \ldots$, when we calculate the mean.)

Thus:

x	f	fx
1	3	3
2	4	8
3	9	27
4	10	40
5	12	60
6	2	12
(sum) = 40		150

Hence mean $\left(= \dfrac{\text{sum of statistics}}{\text{number of statistics}} \right) = \dfrac{150}{40} = 3\cdot75$ years.

In the frequency table, the statistics are already arranged in ascending order. We now have to find where the middle statistic is.

The second column shows there are 40 statistics. Since this is an even number, there is no single middle statistic, and so the median lies between the 20th and the 21st statistics.

We look again at the first two columns:

x	f	
1	3	(3 stats. to here)
2	4	(7 stats. to here)
3	9	(16 stats. to here)
4	10	(26 stats. to here)
5	12	
6	2	

Example continued ➤

Example *continued*

The 20th and the 21st statistics must therefore lie in the 4th Row of the table, and each has the age value 4 years.

$$\text{Hence median} = \frac{4+4}{2} = \frac{8}{2} = 4 \text{ years.}$$

From the table, the mode is 5 years (occurring 12 times).

Cumulative Frequency

In the last Example, we were looking at the *cumulative frequency* as we tried to find the median ('3 stats. to here; 7 stats. to here…').

It is often helpful to add a cumulative frequency column to the table. In this column, the frequency of each row is *accumulated*, as the following illustration shows.

Example

The midday temperature in °C was recorded at a weather station on 41 consecutive days, with the following frequencies:

Temp.	Freq.
11	3
12	8
13	10
14	12
15	8

Add a cumulative frequency column to the table, and use it to determine the median temperature.

(Solution)

Temp.	Freq.	Cum Freq.	
11	3	3	
12	8	11	$(=3+8)$
13	10	21	$(=11+10)$
14	12	33	$(=21+12)$
15	8	41	$(=33+8)$

Since there are 41 statistics, the middle one is the 21st. The third column shows that this is in the third row. Hence median temperature $= 13\,°C$.

The Spread of the Statistics

As we have seen, the mean, median, and mode all give us some idea of a statistical average. More information can be found if we also consider how the statistics of a data set are spread. We can then make comparisons between sets of statistical data.

The Range

> **Remember**
>
> This is the simplest measure of spread.
>
> Range = Largest statistic − Smallest statistic

Example

Find the range of the following set of numbers:

14   3   10   9   4   18   10.

(Solution)   Largest number = 18
Smallest number = 3

Hence range = 18 − 3 = 15.

Example

Find the range of the statistics in the following frequency table:

x	f
21	4
22	5
23	18
24	8

(Solution)   Lowest value = 21, Highest value = 24

Hence range = 24 − 21 = 3.

(Only the first column is of interest here; the frequencies do not matter.)

C Quartiles and Semi-Interquartile Range

The *Median* divides the data set into *two* equal sized groups.
The *Quartiles* divide the data set into *four* equal sized groups.

Key Points

To find the quartiles:
◆ arrange the statistics in ascending order
◆ find the median as before – it is the *second quartile* (Q_2)
◆ now find the middle statistic for each of the lower group and upper group $(Q_1$ and $Q_3)$.

Also, the Semi-Interquartile Range $= \dfrac{Q_3 - Q_1}{2}$.

Example

Find the quartiles and semi-interquartile range for each of the following data sets:

a)   1   3   3   4   6   6   7   10   13   14   14
b)   16   12   24   23   29   30   12   18.

(Solution)

a)   (1   3   <u>3</u>   4   6)   <u>6</u>   (7   10   <u>13</u>   14   14)

$\Rightarrow$ Q_1 Q_2 Q_3

(lower quartile)   (median)   (upper quartile)

Hence semi-interquartile range $= \dfrac{Q_3 - Q_1}{2} = \dfrac{13 - 3}{2} = \dfrac{10}{2} = 5$.

Also, median $(Q_2) = 6$.

b)   16   12   24   23   29   30   12   18.

Rearranging: (12   12 $\uparrow$ 16   18) $\uparrow$ (23   24 $\uparrow$ 29   30)

$\Rightarrow$ Q_1 Q_2 Q_3

$(= \frac{12 + 16}{2})$ $(= \frac{18 + 23}{2})$ $(= \frac{24 + 29}{2})$

Hence $Q_1 = 14$, and $Q_3 = 26{\cdot}5$

$\Rightarrow$ semi-interquartile range $= \dfrac{Q_3 - Q_1}{2} = \dfrac{26{\cdot}5 - 14}{2} = \dfrac{12{\cdot}5}{2} = 6{\cdot}25$.

Also, median $(Q_2) = 20{\cdot}5$.

HOW TO PASS STANDARD GRADE MATHEMATICS

C Standard Deviation

The Range and the Semi-Interquartile Range of the data give us some indication of the spread of the statistics. However, each of these uses only *two* statistics for its calculation.

Standard deviation is a much more reliable measure of the spread of the data, and uses *all* of the statistics in its calculation. It is used often in more advanced work in statistics.

The calculation and interpretation of standard deviation may seem difficult at first, but with a little practice, they become more straightforward.

The larger the value of the standard deviation, the larger is the spread of the data from the mean.

Key Points

The standard deviation s may be calculated from either of the formulae:

$$s = \sqrt{\frac{\Sigma(x - \bar{x})^2}{n - 1}} \quad \text{or} \quad \sqrt{\frac{\Sigma x^2 - (\Sigma x)^2/n}{n - 1}}$$

(where n is the sample size.)

These formulae are printed in the Credit paper. Use the version you are familiar with.

Example

The amount of rainfall (measured in cm) recorded in a highland village in March in each of the last six years was as follows:

$$5 \quad 7 \quad 7 \quad 8 \quad 10 \quad 14$$

Find the mean and standard deviation of these results.

(Solution)   Mean $= \dfrac{5 + 7 + 7 + 8 + 10 + 14}{6} = \dfrac{51}{6} = 8.5$ (cm)

Standard deviation s: (first formula): $\bar{x} = \text{mean} = 8.5$

x	$x - \bar{x}$	$(x - \bar{x})^2$
5	-3.5	12.25
7	-1.5	2.25
7	-1.5	2.25
8	-0.5	0.25
10	1.5	2.25
14	5.5	30.25

$$\Rightarrow \Sigma(x - \bar{x})^2 = 49.50.$$

Hence $s = \sqrt{\dfrac{\Sigma(x - \bar{x})^2}{n - 1}} = \sqrt{\dfrac{49.5}{5}} = \sqrt{9.9} = 3.15$ (to 2 dec. places).

Example *continued* ➢

Example *continued*

Standard deviation s: (second formula)

x	x^2
5	25
7	49
7	49
8	64
10	100
14	196

$$\Rightarrow \Sigma x = 51, \ \Sigma x^2 = 483$$

Hence $s = \sqrt{\dfrac{\Sigma x^2 - (\Sigma x)^2/n}{n-1}} = \sqrt{\dfrac{483 - (51)^2/6}{5}} = \sqrt{\dfrac{483 - 433\cdot5}{5}}$

$= \sqrt{\dfrac{49\cdot5}{5}} = \sqrt{9\cdot9} = 3\cdot15$ (as before).

Statistical Comparisons

In some questions, you may be asked to comment on the differences between two sets of data.

Example

The mean mark in a fourth year class test is 73·4. John's test mark is 72, and Anna's is 95. Comment on these marks.

(Solution)
John's mark is only slightly below the class mean but Anna's is well above the mean.

Example

A bus company records the times taken (in minutes) for the journey from the city centre to the bus depot. The mean and standard deviation of the journey times are as follows:

	Mean	Standard deviation
Weekdays	34	2·5
Weekends	27	3·8

Make two valid comparisons between the two sets of data.

(Solution)
The mean journey time at the weekend is less than that on weekdays.
There is, however, more variation in the journey times at the weekend (since the standard deviation is greater).

For Practice

(The answers to the following questions are given in Appendix 2.)

18.1 The following frequency table shows the ages of the members of the junior section of a tennis club:

age	freq
10	2
11	4
12	3
13	5
14	3
15	4
16	2

Find the mean age, the median age, and the range of age.

18.2 The numbers of goals scored in a season by the clubs in a football league were as follows:

18   26   39   23   21   17   12   27   33.

What is the range, median, and mean of this data?

18.3 What is the semi-interquartile range of the data in Question **18.2**?

18.4 The weights of six dogs of the same breed were measured to the nearest kilogram at a dog show. The results were as follows:

12   15   14   15   18   16.

What was the mean weight and the standard deviation of the weights?

HOW TO PASS STANDARD GRADE MATHEMATICS

Chapter 19

STATISTICAL DIAGRAMS AND PROBABILITY

Statistical information is frequently represented in a diagram. This can be very helpful or very misleading. We need to be able to interpret information correctly from a diagram so that we may come to a valid conclusion. Some statistical diagrams are helpful, too, in allowing us to calculate the probability of an outcome or event.

What You Should Know

a) for General Level: how to
- ◆ extract data from pie charts, scattergraphs, stem and leaf diagrams
- ◆ construct scattergraphs, stem and leaf diagrams, line graphs
- ◆ draw the best fitting line on a scattergraph
- ◆ state the probability of an outcome.

b) and in addition, for Credit Level: how to
- ◆ extract data from box plots, dot plots
- ◆ construct pie charts, box plots, dot plots
- ◆ find the equation of the line of best fit for a scattergraph
- ◆ calculate probability on a scale between 0 and 1.

Statistical Diagrams

The Pie Chart

The pie chart presents the information as a circular 'pie' or cake. Slices or sectors of the pie are then used to represent sections of the information. The angle of the sector is proportional to the size of the quantity the sector represents.

Example

A survey was made of the destinations of 200 school leavers. The results are shown in the following pie chart.

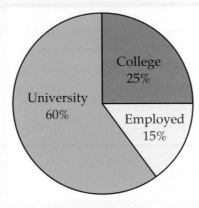

How many students does each section represent?

(Solution)

University $60\% = \dfrac{60}{100} \Rightarrow$ number of students $= \dfrac{60}{100} \times 200 = 120$

College $25\% = \dfrac{25}{100} \Rightarrow$ number of students $= \dfrac{25}{100} \times 200 = 50$

Employment $15\% = \dfrac{15}{100} \Rightarrow$ number of students $= \dfrac{15}{100} \times 200 = 30.$

(Check: $120 + 50 + 30 = 200$).

The Scattergraph

This is a useful method of displaying and comparing two sets of data which are related in some way. The scattergraph is simply a collection of points on a coordinate diagram.

The points do not all lie on a straight line, but a line of *best fit* can be drawn which passes through some of the points, and is close to most of the others.

If you are drawing a scattergraph, you should try to have roughly the same number of points above and below the line.

Example

The following scattergraph shows the results of two tests taken by the same 15 pupils. The tests were in Physics and Maths. The line of best fit has been drawn.

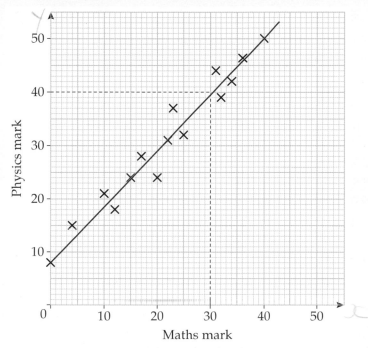

A sixteenth pupil missed the Physics test, but scored 30 marks in the Maths test. What would be the best estimate of his likely Physics mark?

(Solution)
By projecting a Maths score of '30' on to the line of best fit, we see that his likely Physics score is '40'.

In a question like this, you may be asked to find the *equation* of the line of best fit. (Remember $y = mx + c$, where m is the gradient of the line, and c is the intercept on the y-axis.)

Here, the intercept is the point $(0, 8)$. The gradient is calculated from two well-spaced points on the line such as $(0, 8)$, and $(40, 50)$.

$$\text{Gradient} = \frac{\text{vertical}}{\text{horizontal}} = \frac{50 - 8}{40 - 0} = \frac{42}{40} = \frac{21}{20}.$$

The equation of the line of best fit is therefore $y = \frac{21}{20}x + 8$.

The Box Plot

A box plot is a very convenient way of illustrating the smallest and largest statistics, and also the quartiles. It provides a five-figure summary of the data.

Example

Golf tees are sold in packets of '50'. However, a sample of 11 packets contained the following numbers of tees:

52   58   56   55   48   56   58   50   57   49   51.

Find the median and upper and lower quartiles for this data and construct a box plot to illustrate the data.

(Solution)
(arranging the data in order):

48   49   50   51   52   (55)   56   56   57   58   58

⇒   Q1   Q2   Q3

The Median is therefore 55 (tees), the lower quartile is 50 (tees), and the upper quartile is 57 (tees). The box plot then appears as follows:

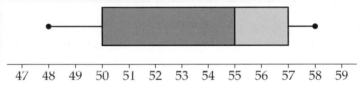

47   48   49   50   51   52   53   54   55   56   57   58   59

The Dot Plot

A dot plot gives us a simple technique to illustrate small numbers of data. The dot plot is really just a simple form of bar graph, and its value lies in showing the *shape* of a distribution of data.

Example

Mr Cummings takes a note of the number of telephone calls he receives each day over a three-week period. They are as follows:

Calls each day	0	1	2	3	4	5	6
Number	2	3	6	3	2	4	1

Show these results on a dot plot.

Example continued ➤

Example continued

(Solution)

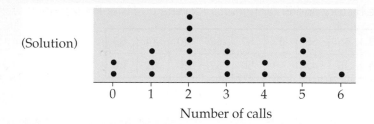

Number of calls

The Stem and Leaf Diagram

The stem and leaf diagram provides us with a very simple but effective way of displaying all of the data in a numerical ascending order.

If our data is in the form of two digit numbers, then the first digit becomes part of the 'stem', and the second digit becomes a 'leaf'.

The diagram must include a key which explains the arrangement of stem and leaf. It is usually helpful to have the data in ascending order before constructing the stem and leaf.

Example

The following midday temperatures (in °F) were recorded at the Glasgow Weather Centre in the first thirteen days of May.

   43   49   50   50   54   57   58   58   62   66   71   74   75.

Illustrate these temperatures in an ordered stem and leaf diagram.

(Solution)

```
4 | 3   9
5 | 0   0   4   7   8   8
6 | 2   6
7 | 1   4   5
```

($n = 13$;   7|1 represents 71).

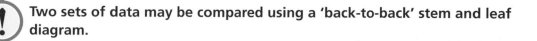

Two sets of data may be compared using a 'back-to-back' stem and leaf diagram.

Example

The table shows the History and English marks for nine pupils at Prelim. time.

History (%)	53	55	58	61	62	62	70	74	83
English (%)	34	36	42	49	54	57	60	65	66

Illustrate these results in an ordered back-to-back stem and leaf diagram.

(Solution)

```
      6   4 │ 3 │
      9   2 │ 4 │
      7   4 │ 5 │ 3   5   8
   6  5   0 │ 6 │ 1   2   2
          │ 7 │ 0   4
          │ 8 │ 3
```

($n = 18$ 6|2 represents 62).

Probability

What is the probability that it will be sunny tomorrow?
What is the probability that we will win on Saturday?

Key Points *and* Definitions

In Mathematics, *probability* is a numerical measure of an event happening.
It is defined as

$$\text{Probability} = \frac{\text{number of favourable outcomes}}{\text{number of possible outcomes}}.$$

Example

A letter is chosen at random from the word SUCCESS.
Find the probability of   **a)** an E   **b)** an S   **c)** a T.

(Solution)

a) (since only one E) $p(E) = \dfrac{1}{7}$

b) (since three Ss) $p(S) = \dfrac{3}{7}$

c) (since no T) $p(T) = \dfrac{0}{7} = 0.$

! **A probability must always have a value from 0 and 1.**

Example

Two similar dice are rolled at the same time. What is the probability that the total score is greater than 9?

(Solution)

The table shows all possible total score outcomes:

[Die (1)]		1	2	3	4	5	6	
[Die (2)]	1	2	3	4	5	6	7	
	2	3	4	5	6	7	8	
	3	4	5	6	7	8	9	(total score)
	4	5	6	7	8	9	**10**	
	5	6	7	8	9	**10**	**11**	
	6	7	8	9	**10**	**11**	**12**	

Six scores out of 36 are greater than nine.

Hence probability $\left(= \dfrac{\text{number of favourable outcomes}}{\text{number of possible outcomes}}\right) = \dfrac{6}{36} = \dfrac{1}{6}$.

For Practice

(The answers to the following questions are given in Appendix 2.)

19.1 The following dot plot shows the speeds in miles per hour of 20 cars as they entered a 30 m.p.h. restriction zone.

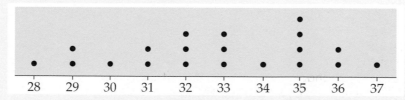

Using this information, find the mean speed and the median speed.

19.2 A die is rolled, and a coin is tossed.

a) Construct a table to show all possible outcomes.

b) What is the probability of obtaining a 'Head' **and** a score greater than 4?

NON-CALCULATOR PRACTICE QUESTIONS (GENERAL)

Exercise 1

1   a) $16{\cdot}2 + 5{\cdot}78 - 9{\cdot}4$

     b) $8{\cdot}37 \times 40$

     c) $0{\cdot}276 \div 6$

     d) $\dfrac{2}{3}$ of $198\,\text{g}$.

Exercise 2

1   a) $10{\cdot}21 - 3{\cdot}754$

     b) $9{\cdot}67 \times 8$

     c) $388 \div 400$

     d) $5 \times 1\dfrac{3}{8}$.

Exercise 3

1   a) $26{\cdot}57 - 4{\cdot}8 + 1{\cdot}39$

     b) $0{\cdot}94 \times 70$

     c) 60% of £210

     d) $\dfrac{3}{4} + \dfrac{5}{12}$.

Exercise 4

1   a) $4 - 2{\cdot}73 + 5{\cdot}68$

     b) 26×12

     c) $534 \div 300$

     d) 40% of $25\,\text{kg}$.

Exercise 5

1   a) $34{\cdot}5 - 27{\cdot}83$

     b) $6{\cdot}19 \times 3000$

     c) $224 \div 70$

     d) 15% of $400\,\text{ml}$.

APPENDIX 2

ANSWERS

Chapter 3

3.1	34·2 litres
3.2	11·2%
3.3	1050
3.4	$\dfrac{5}{6}$
3.5	$\dfrac{7}{16}$
3.6	$1\dfrac{8}{21}$
3.7	$\dfrac{5}{8}$
3.8	$3\dfrac{3}{5}$
3.9	0·714
3.10	$\dfrac{7}{10}$, 0·707, 0·71, $\dfrac{5}{7}$, 73%
3.11	19°
3.12	−13
3.13	21
3.14	3
3.15	−2.

Chapter 4

4.1	8 510 000
4.2	$3·07 \times 10^9$
4.3	$2·32 \times 10^7$
4.4	$3·65 \times 10^4$
4.5	$4·72 \times 10^{12}$.

Chapter 5

5.1	104		
5.2	35 litres		
5.3	12 m		
5.4	6 min		
5.5	a) $P \propto \dfrac{X}{\sqrt{Y}}$	b)	P is halved
5.6	a) 12 m		
	b) $k = \dfrac{6}{5}, D = \dfrac{6}{5} \times T$.		

Chapter 6

6.1	12 h 45 m
6.2	44·3 m.p.h.
6.3	1·987 m, 1·973 m.

Chapter 7

7.1	a) £720	b)	£18 720
7.2	$7\dfrac{1}{2}$ hours		
7.3	a) £55	b)	22·9%
7.4	a) £1036	b)	£141
7.5	a) £94	b)	£92·40
7.6	buildings (£72·91), contents (£162·54), Total £235·45		
7.7	£255·15		
7.8	£56 each, £112 total.		

Chapter 8

8.1	a) 80 p	b)	7 min
8.2	a) €923	b)	279 sf
8.3	£1660		
8.4	£179·17		
8.5	£3629·16		
8.6	518·4 mg.		

Chapter 10

10.1	24 m²
10.2	1200 cm².

Chapter 11

11.1	4973·6 cm²
11.2	7·3 m
11.3	41·3°
11.4	7·1 cm².

Chapter 12

12.1	7·8 m
12.2	1·97 m
12.3	b) 10·3
12.4	yes ($24^2 + 7^2 = 25^2$).

Chapter 13

13.1	440 cm²
13.2	100·5 cm³
13.3	120 cm³
13.4	631·4 cm

13.5   a) $4800\,\text{cm}^2$ b)   2

13.6   a) $1400\,\text{cm}^3$ b)   5 cm.

Chapter 14

14.1 $5a + 4$

14.2 $x^3 + 3x^2 - x - 3$

14.3 $2b(2a - 5c)$

14.4   a) $(p + 5q)(p - 5q)$

        b) $(2x + 1)(x - 4)$

14.5 $x = 10$

14.6 $x = 3$

14.7 $x = 1, y = -2$

14.8   a) $x = 5, -5$ b) $x = -\dfrac{1}{3}, 2$

14.9 $x = 1\cdot35$ or $-1\cdot85$.

Chapter 15

15.1 $12\cdot5$

15.2 $a = \dfrac{v^2 - u^2}{2s}$

15.3   a)   17   b) $t = \pm7$

15.4 $\dfrac{x + 1}{x + 2}$

15.5   a) $\dfrac{4 + x^2}{2x}$ b) $\dfrac{6}{pq}$

15.6 $a - 1$

15.7   a) $3\sqrt{7}$ b) $\dfrac{5\sqrt{3}}{3}$ c) $3\sqrt{2}$.

Chapter 16

16.1 $h = 78\cdot14$ cm

16.2   angle BCA $= 38\cdot8°$

         angle BAC $= 71\cdot2°$

         BC $= 4\cdot53$ cm

16.3   a) $183\cdot7$ m   b) $1069\,\text{m}^2$.

Chapter 17

17.1

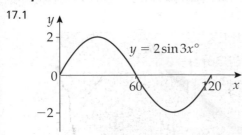

17.2 $x = 41\cdot8°, 138\cdot2°$.

Chapter 18

18.1   mean $= 13$, median $= 13$, range $= 6$

18.2   range $= 27$, median $= 23$, mean $= 24$

18.3   siqr $= 6\cdot25$

18.4   mean $= 15$ (kg) st dev $= 2$ (kg).

Chapter 19

19.1 $32\cdot8$ m.p.h., 33 m.p.h.

19.2   a)   1h 2h 3h 4h 5h 6h

            1t 2t 3t 4t 5t 6t

        b) $p = \dfrac{1}{6}$

Appendix 1

Exercise 1

1   a) $12\cdot58$ b) $334\cdot8$

   c) $0\cdot046$ d)   132 g.

Exercise 2

1   a) $6\cdot456$ b) $77\cdot36$

   c) $0\cdot97$ d) $\dfrac{55}{8}$.

Exercise 3

1   a) $23\cdot16$ b) $65\cdot8$

   c)   126   d) $1\dfrac{1}{6}$.

Exercise 4

1   a) $6\cdot95$ b)   312

   c) $1\cdot78$ d)   10 kg.

Exercise 5

1   a) $6\cdot67$ b)   18 570

   c) $3\cdot2$ d)   60 ml.